Mechanic on the Wing

MECHANIC
ON THE
WING

*The Untold Story of Carrier Aircraft
Service Unit Eleven
(CASU-11) 1943 - 1946*

William Henry Little

MECHANIC ON THE WING

Cover photos: Top left is a Navy photo of the first aircraft to land at Yontan Airfield, Okinawa from the 58th Naval Construction Battalion Cruise Book 1942-1945. Just below is a blow-up detailing a wing rider guiding the pilot to ensure the plane stays on a safe path. Top right is a photo of the Asiatic-Pacific Campaign Medal. Bottom right is a Marine Corps photo with another wing rider guiding his pilot to a safe parking spot.

Andrea Sikkink of As You Wish Arts created the beautiful cover for this book.

Dedication

This book is dedicated to

Harry Absolum Hays, Jr.

Juan Leal, Jr.

Robert Henry Little

John Joseph McAteer

Patrick Gregory O'Flynn

Henry Franklin Parker

Durward West

David Anderson Wilson

These eight men have spoken to me across the years, through their voices, their writings, their photos, and by the stories shared by their sons and daughters. This convinced me that a book about CASU-11 was not just warranted, it was mandatory.

War talk by men who have been in a war is always interesting.

Mark Twain,
Life on the Mississippi

Table of Contents

Introduction

It was just a box. A simple, light brown box about the size of a small cigar box, given to me by my brother at the passing of our father. A label on one side read, "NAVY DAYS DIARY - 1 YEAR DIARY - LOCK & KEY - 1141 LEATHERETTE BLUE." On the inside cover of the diary within was written, "To Robert, Love, From your cousin, Esther, 1942."

There were no entries in the diary. I wish my dad, Aviation Machinist's Mate First Class Robert Henry Little, had scribbled a few words in this thoughtful gift. Like many World War II veterans, my dad rarely talked about his experiences in the war. Sadly, like many children of World War II veterans, I neglected to ask him about the war, and then it was too late. He died in 2003.

The empty diary had piqued my curiosity. I knew that my father had served in Carrier Aircraft Service Unit 11 (CASU-11) in Guadalcanal. Other than the few stories he shared with me, I knew very little of this extraordinary time in his life. Beginning with my father's military service record, I outlined a rough chronology of CASU-11. I also began searching for CASU-11's War Diary. A War Diary was an official document required by the Chief of Naval Operations from each Navy command. Unfortunately, the war diary for CASU-11 was either lost or never written.

Introduction

Luckily, the personal diary of Aviation Metalsmith Chief Harry Hays was made available to me. Additionally, I was able to interview Motor Machinist Chief Patrick O'Flynn who also served on Guadalcanal with CASU-11. Finally, the War Summary Reports issued by the Pacific Fleet Staff of Admiral Nimitz and the many documents on the website FOLD3 were extremely helpful.

As a result, the idea for a book began to take shape that would include not only my father's time in CASU-11 but also a complete history of the unit.

During World War II, almost 70 United States Navy Carrier Aircraft Service Units (CASUs) were formed and deployed to support naval aircraft. Faced with a lack of aircraft carriers early in the war and no organizational means to support quickly constructed airfields on islands in the Western Pacific, the CASUs were a wartime necessity.

CASU-11 was established on Friday, 22 January 1943 and decommissioned on Friday, 1 November 1946. This was a total of 1,379 days, or three years, nine months, and ten days of service. During that time CASU-11 traveled approximately 27,000 miles and more than 1,700 men passed through its ranks. Finally, it must be noted that CASU-11 bookends World War II in the Pacific. They were present on Guadalcanal, the first island reclaimed from Japanese occupation, and they were present on Okinawa, the last island taken before the war ended.

Part One - San Diego, California

On 1 July 1941, Robert Henry Little, the man who would become my dad, registered for the draft with the Hartford County Board No. 1 in Bel Air, Maryland. The 21-year old dump truck driver made repeated attempts to enlist in the U.S. Navy. He was rejected each time. As a young boy, he had rheumatic fever and suffered heart damage. An easily detected heart murmur lead to his rejection at each of his Navy physicals. Finally, on 13 January 1942, he raised his hand and was sworn into the United States Navy.

How he managed to finally pass the physical is questionable. According to family lore, Little visited the family doctor and received a medicine that eliminated the murmur. With the bombing of Pearl Harbor the month before and the United States entry into World War II, he may have just been lucky. There was a significantly increased need for military personnel.

Little shipped off to Navy Training Station, Norfolk, Virginia, for five weeks of boot camp. This was followed by six months of training at Aviation Machinist Mate's school. Then he completed four months of additional training in Aircraft Fuselage, Hydraulics, Fuel Systems and Controls at the Advanced Base Aviation Training Unit, Norfolk, Virginia.

On 16 September 1942, Little became an Aviation Machinist Mate (AMM3c). In military parlance this was his rating or skill set. He would be responsible for the maintenance of naval aircraft. This included cleaning, fueling, servicing, and making emergency repairs on all types of airplanes. It was a critical rating for the Carrier Aircraft Service Unit (CASU). The "3c" in his rating indicated his rank or pay grade; Little was now a Petty Officer Third Class.

Almost a year after he enlisted, and shortly after Christmas of 1942, Little started his 2,600-mile trip west to Naval Air Station (NAS) San Diego, California, and CASU-5. Leaving his parent's farm in Hartford County on 1 January 1943, he caught a westbound train in Baltimore and arrived in San Diego on 7 January.

Meanwhile from points all over the 48 states, 500 other Sailors were traveling to NAS San Diego and CASU-5. None of them knew that CASU-5 was just a temporary stop, a collection unit permanently stationed at NAS San Diego. At CASU-5, Sailors gathered and trained for later assignments to the new CASU's that were being rapidly organized and shipped out. This would have been the first time future members of CASU-11 would have met. Members like Motor Machinist Mate First Class Patrick Gregory O'Flynn from Mississippi would work on vehicles in the CASU-11 motor pool. Aviation Chief Metalsmith Harry Absolum Hays, Jr. born in Indiana but living in California, and Aviation Chief Metalsmith David Anderson Wilson of Massachusetts and Aviation Metalsmith Third Class Henry Franklin Parker born in Kansas and currently living in California, were all specialists in airplane structures. These men knew metalsmith methods and materials and were also critical ratings for a CASU. Aviation Radioman Third Class Durward West from Illinois would work on the communications equipment. Ship's Cook Third Class (SC3c) Juan Leal, Jr. of Texas would serve as one of the cooks.

Train rides to San Diego were laborious in those days, as cargo had priority over passenger trains. O'Flynn remembered how long the trip was from Mississippi. "Three days in those days," he said. "Cause every time you got on a train, if something important was coming through, you go on a side line to let traffic go."

None of the Sailors knew where they would be assigned, when they might go overseas, or where in the Pacific they might be sent. In fact, most of these men were taking their first train ride and were farther from their homes than they had ever been in their lives.

Men holding or receiving commissions as officers in the United States Navy were also in training for duty with CASU-11. One of these officers was Lt. John J. McAteer, who was attending the Navy Supply School at Harvard University in January. He would later graduate on 2 April 1943 before joining CASU-11 at Henderson Field, Guadalcanal in May 1943.

On 30 December 1942, an aviation qualified officer, Lt. Cmdr. Isaac Schlossbach, arrived in San Diego and checked in at CASU-5 with orders to become the first commanding officer for CASU-11. (Appendix C provides details about all of the Commanding Officers of CASU-11.) By mid-January, more than 20 officers and 500 Sailors, including AMM3c Little, had reported to CASU-5 for further duty with CASU-11. Few of these men knew their time in San Diego was almost over. On 22 January 1943, Lt. Cmdr. Schlossbach formally commissioned CASU-11 as an official unit of the U.S. Navy.

Document from Lt. Cmdr. Schlossbach's Navy Service Record.

"Schlossbach ... it was hard to pronounce," recalled O'Flynn, "I never saw him but once, I don't think. He was the only lieutenant commander but still can't tell much about him. He only had one eye. We called him one-eyed Schlossbach. You know we were all very young, and a lieutenant commander was a big shot."

Not a word was spoken during the commissioning ceremony about when and where CASU-11 was headed. Many hoped it would not be Guadalcanal, the place repeatedly appearing in the headlines of San Diego's newspapers. The Marines were already there, and they were having a tough time with the Japanese and the tropical jungle.

The military is known for its seemingly endless cycles of "hurry up and wait." And now that CASU-11 officially existed, its men had entered what would be one of many periods of waiting.

San Diego was neither the worst nor the best place to wait. Men were housed in old aircraft hangers, quickly built, metal-roofed Quonset huts, or tents. They had no heat, electricity or lockers. Their quarters were over-crowded. According to a letter written by the commanding officer of CASU-5, "Into the space capable of handling approximately 250 men, more than 600 men were housed."[1] Bunk beds filled every bit of floor space. In some places, the beds were placed on bricks which permitted a third person to sleep on a mattress pad under the bottom bunk. Later Sailors were assigned to shift work so that day workers could use the bunk at night and night workers could use them during the day. This was called hot bunking.

During the last half of 1942, the CASU-5 War Diary reported, "At this time there was only one mess hall available to accommodate the needs of the entire sea and air personnel based at this station. The situation of feeding of crew was extremely acute and very unsatisfactory, as the hours worked were from very early morning until past midnight of each day."[2]

Many man-hours were lost due to inadequate transportation. The men did a lot of walking, sometimes with a sea bag on their back. They walked to get to work, to get to chow and to get to their bunk. The distance from the barracks to work could be as much as two miles, then one mile to eat lunch, one mile back to work followed by two miles back to the barracks at the end of the day. The day concluded with another mile from the barracks to the evening meal and a final mile back to their bunk.

Not surprisingly, San Diego grew bigger and bigger due to the rapidly increasing Navy population. Hundreds of military service personnel arrived on each passenger train, with multiple trains arriving each day. Between 1940 and 1946, San Diego's population increased by 60 percent. San Diego would never again be a "sleepy Navy town,"[3] as Gerald Nash had described its condition before the war.

Finally, on 1 February 1943, CASU-11 received word that their ship was coming, and the men made preparations to board. Everyone washed their clothes, repacked their sea bags, cleaned their rifles, yes, every Sailor had their own rifle, and sent letters home. Except for Lt. Cmdr. Schlossbach and the Ship's Master, nobody knew for sure where they were headed. In those days the commanding officer was told their destination and handed a sealed, secret orders package to be opened after the ship was at sea. Upon its opening, sometimes all officers and Sailors would be immediately informed, sometimes only the officers, and sometimes the commanding officer would keep it to himself.

San Diego, California

Part Two - Espiritu Santo, New Hebrides

"We went to Espiritu Santo first; that's where the ship was going that took us over there," O'Flynn said during an interview. He continued, "I knew then that we were going to Guadalcanal." Hays wrote in his diary the day CASU-11 departed Espiritu Santo, "Scuttlebutt is Guadalcanal. Am praying we make it." Based upon these two anecdotal statements, it appears all hands were told after leaving San Diego that the next stop was Espiritu Santo, and that shortly before leaving Espiritu Santo, CASU-11 was advised the next stop was Guadalcanal. Espiritu Santo is located 15 degrees south of the equator in the New Hebrides archipelago which bounds the eastern edge of the Coral Sea.

However, as the Sailors of CASU-11 walked down the pier in San Diego they did not know their final destination. They also did not know the ship's name that they would ride until they read the name painted on a canvas banner attached to the gangway.

The name of the ship that took them across the Pacific was not included in the service records of the members of CASU-11. The only evidence that CASU-11 even went to sea, were two stamped entries in every service record stating:

"CROSSED THE EQUATOR
QUALIFIED SHELLBACK
2-16-43"

"Qualified GOLDEN DRAGON
Crossed 180th Mer.
February 21, 1943"

The first was crossing the equator and the second was crossing the 180th meridian known as the International Date Line.

While I was doing research for this book, I found the document shown below in my father's personal scrapbook. It was a folded document the size of a business card with the words "Domain of Neptunus Rex," meaning "Domain of King Neptune," on it. Since this card was surrounded with 1950s paraphernalia and under a plastic sleeve, I had passed over it many times. By chance, I decided to take a closer look.

To my surprise, opening the card revealed that CASU-11 was on board the SS President Polk when it crossed the Equator on 16 February 1943. This was the same date as the stamped "CROSSED THE EQUATOR." entry in my father's service record. A similar document from Hays' personal records verified CASU-11's presence on board the SS President Polk when they traveled across the Pacific Ocean.

Crossing Equator Card Front Cover. From Robert Little.

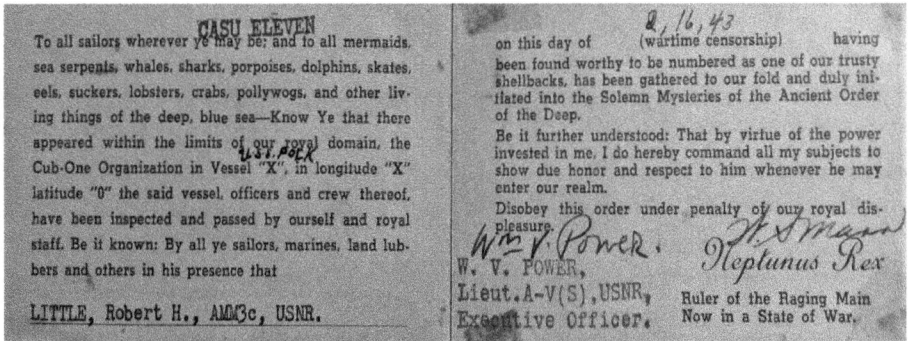

To all sailors wherever ye may be; and to all mermaids, sea serpents, whales, sharks, porpoises, dolphins, skates, eels, suckers, lobsters, crabs, pollywogs, and other living things of the deep, blue sea—Know Ye that there appeared within the limits of our royal domain, the Cub-One Organization in Vessel "X", in longitude "X" latitude "0" the said vessel, officers and crew thereof, have been inspected and passed by ourself and royal staff. Be it known: By all ye sailors, marines, land lubbers and others in his presence that

LITTLE, Robert H., AMM3c, USNR.

on this day of (wartime censorship) having been found worthy to be numbered as one of our trusty shellbacks, has been gathered to our fold and duly initiated into the Solemn Mysteries of the Ancient Order of the Deep.

Be it further understood: That by virtue of the power invested in me, I do hereby command all my subjects to show due honor and respect to him whenever he may enter our realm.

Disobey this order under penalty of our royal displeasure.

W. V. POWER,
Lieut.A-V(S),USNR,
Executive Officer.

Neptunus Rex
Ruler of the Raging Main
Now in a State of War.

Inside crossing equator card. From Robert Little.

Prior to World War II the SS President Polk was owned by the American President Lines. The San Francisco Maritime National Park Research Center holds the archives of this shipping company and their deck logs.[1] During World War II ship movements were classified. Consequently, ports of call were left unidentified and only logged as Port 1, Port 2, and so forth.

On Monday, 1 February 1943, the SS President Polk was pier side in San Francisco when she received orders from the Department of the Navy to proceed to San Diego. For this cruise, San Francisco was Port 1 and San Diego was Port 2. By mid-morning on Friday, 5 February, the SS President Polk was all fast and secured by mooring lines to the south side of the Navy Pier at the foot of Broadway in San Diego, California. Today this is the location of the USS Midway Museum on the San Diego waterfront.

In addition to the SS President Polk, hundreds of other unarmed, civilian ships were recruited to handle the logistical requirements of World War II. These ships would be travelling in very dangerous waters to deliver required cargo, fuel and personnel. Ship protection from enemy attack especially Japanese submarines and aircraft was essential. Consequently, after coming into Navy service, each ship was equipped with Navy guns designed to defend against these threats. The U.S. Navy Armed Guard was onboard to provide manpower for maintaining and operating these weapons. Armed Guard officers had the task of maintaining naval discipline, creating an effective gun crew and establishing a working relationship with the ship's master and merchant crew.

Civilian stevedores loaded cargo all day Saturday and Sunday and again Monday morning. This quick cargo-loading effort, versus what will become known as "combat loading," would have a negative impact upon the military passengers the SS President Polk carried to the South Pacific. On Monday, 8 February, at 1300, the embarkation of troops onto the SS President Polk commenced and by noon on Tuesday, 9 February, everyone was on board. Additional information on the SS President Polk is provided in Appendix E along with pictures and characteristics of the other ships that transported CASU-11 back and forth across the Pacific.

From February 1943 through August 1944 Hays kept a daily journal telling his story of life with CASU-11. On 28 March 2016, Hays's daughter, Melissa Hays, graciously began providing pages of Harry's diary to me. The picture below was attached to an email from Melissa which brought tears to my eyes. I had already read the log entries for the SS President Polk departing San Diego - on 9 February 1943 Harry's words almost matched the deck log for the SS President Polk. His entire 17-month dairy will be provided chronologically as the story of CASU-11 unfolds.

Page 1. Harry Hays' Journal.

FEBRUARY 1943

Left Tues. 9th 2:00 P.M. Circled harbor until about 5:30 P.M. Shoved off. Plenty rough. 1000 Marine Raiders, about 500 Marine Aircraft Crew, 700 Sailors. Almost all sick. Not me. Living quarters and conveniences very bad. Sailing West & South. Slept pretty well. From the diary of Harry Hays, February 1943.

At 1303 on 9 February, the SS President Polk's deck log reported that all lines were cast off from the pier in San Diego. The SS President Polk then maneuvered to calibrate degaussing coils and adjust the ship's compass by

13

passing through the degaussing range and circling around the harbor swinging the compass. Degaussing coils are electric coils installed inside a ship that can be adjusted to reduce the ship's magnetic field and, consequently, reduce the ship's susceptibility to explosive mines. At 1830 the compass adjustment was complete. At 1842 the SS President Polk departed San Diego Harbor. The engines were ordered to full ahead and the SS President Polk sailed away at a mere 16.5 knots or 19 miles per hour.

During the February to March Pacific transit, the SS President Polk's crew consisted of Merchant Marine personnel, Naval Armed Guards, a Communications Squadron and a Hospital Corpsman. The Merchant Marine crew that operated the ship consisted of 345 officers, engineers and deck hands. The average Armed Guard detachment had 27 men, including six petty officers, a coxswain, two gunner's mates, two signalmen and a radio operator. On larger crews, a boatswain would also be included in the ranks.

The War Diary dated February 1943 from the Naval Supply Depot, San Diego reported, "The SS Polk embarked a total of 1,936 personnel on 9 February 1943."[2] The exact number of CASU-11 personnel onboard is vague, but estimates have the number close to 500 men. The first solid evidence on CASU-11's manning numbers was a document dated 1 September 1943, well after CASU-11's arrival on Guadalcanal, and it reported CASU-11 as having 22 officers and 550 Sailors.

Most of the 1,936 men had never been to sea before, and many got seasick soon after departure. It was crowded; the SS President Polk was originally designed to hold about 90 passengers and was hastily converted with bunks to transport Sailors and other military personnel. On the first night, Hays probably heard a cacophony of wheezing and hissing, coughing fits, retching, belching, farting, snoring in several pitches, soft moaning, sobbing and cursing. These noises emerged from hundreds of bodies and fused into a single unending human murmur somewhat muted by the background shushing of ocean water rushing by the hull, and the mechanical whir from electric generators and other equipment. What follows are three more personal observations about this cruise.

Bum Phillips, in his 2010 book, *Bum Phillips: Coach, Cowboy, Christian*, reported that he was on board the SS President Polk with the Fourth Marine Raiders for this same cruise. He wrote, "I slept below deck in rooms with cots stacked five high. The space between each cot left just enough room for a Marine's nose to point upward. The transport lacked air conditioning. Poor ventilation to the lower decks made us sweat and stink, and we were angry. It was crowded, and it took forever to get there. We were like sardines in a can."[3]

"The Polk did not have the same facilities as the Navy's APAs or AKAs [naval transport ships]," wrote Daniel Z. Marsh, a Marine from VMD-154, a Marine Corps aerial reconnaissance /photographic squadron. "Consequently, they could provide only two meals a day plus a dry bread sandwich for lunch served topside. However, we did not let that take the edge off the excitement of being at sea and headed for the war zone. The ship was running without escort employing evasive action that caused us to wonder a bit but not for long. Most of us were land lubbers and thoroughly enjoyed the new adventure of being at sea. When we crossed the equator and the International Date Line, we quietly endured the usual initiation festivities. It was rather neat to go to bed one night and wake up yesterday. When we entered the South Pacific we marveled at the deep blue color of the water and how at times it was smooth as glass. Sixteen days passed by, before we entered the harbor at our destination. There we crowded the rails to gaze at the dark, brooding hulk of our new home and wondered."[4]

Aviation Machinist Mate First Class John Rigsby provided his own description of the conditions on board the SS President Polk, "Well the lower 'tween decks area where we were, they packed it five bunks high. The bunks were about 18 inches apart. If you laid on your side, your shoulders would be rubbing on the bottom of the guy above you, so you didn't often do that. You learned to just sleep on your back and try not to snore. And the gear you had got dumped in the aisle between the bunks; the poor guys on the bottom couldn't get in or out around them. We must have had a thousand men in that hold. It was just so jammed, and all you had to get topside was a wooden ladder, stairs, you know, like the kind of stairs we might have in our home basement or something like that. So I could just imagine if we had a problem, like a torpedo attack or a shelling, your chances of getting up those stairs, getting up topside where you might have a chance of survival was out of the question. You might be one of the lucky to get up the stairs, but it was just not going to work. That was the SS President Polk, and we managed to get to Espiritu Santo, New Hebrides without any problem."[5]

Wed. 10th Chow bad. No convoy as yet. Maybe none at all. Had 12 to 4 P.M. watch. Got towel from Sea Bag. Calox spilled. Must have Life Jackets at all times. Had alert drill. Hope we never need it. Haven't shaved yet. Sailors and Marines played for us. Pretty good. Sax and guitars. Haven't gambled yet. From the diary of Harry Hays, February 1943.

The SS President Polk sailed alone with no destroyer protection to the island of Espiritu Santo, New Hebrides. This island inspired James Michener's book "Tales of the South Pacific," which later was turned into the Rogers and Hammerstein musical "South Pacific." The distance between San Diego and Espiritu Santo is 5,207 nautical miles, which the SS President Polk traversed in 16 days. Calox was a brand of tooth powder packaged in a can. With sea bags stuffed in the aisle between bunks, one can imagine why retrieving a towel was an accomplishment worthy of a diary notation.

Thurs. 11th Nothing new. Still haven't shaved. No hot water. Have to use helmet to get fresh water. Had exercise on deck and another alert. Chief gave me some books to study. Survivor gave talk. Got to canteen for first time. Fri. 12th Got up early and shaved. Still no shower. Nothing new. Sat 13th Hot. Got lecture on camouflage, adjusted Gas Mask. Another alert. Can sleep on deck if we want. Sure is monotonous. Might make 1st class. Had discussion on metalwork. Will shave tomorrow A.M. if I can possibly do it. Saw Flying Fish for first time today. Sun. 14th Showered Hurrah! Also washed socks, skivvies, towel, shirt. Had lecture on what to do when we abandon ship. Sharks, life-rafts, first aid, etc. Went to Mass. Might be fights tonight. Hottest day yet. From the diary of Harry Hays, February 1943.

Helmets were the only available water-tight containers and Sailors and Marines routinely used them to carry fresh water back to their bunks. The men could also visit the canteen to purchase soda pop and candy bars with the additional cost of waiting in a long line. Many of the men chose to sleep on deck to escape the heat of the berthing, which sometimes reached 100 degrees F or more. The "fights" Hays mentioned are boxing matches, a regular recreation activity within the Navy at that time. Little had a scar on his eyelid from his participation in this sport.

Mon. 15th Boy oh boy is it hot. So crowded you can't find a place to sit down. Sure looking forward to the time when I'll be comfortable again. Have to wait until April 1st to make 1st class. Not enough time in. Rained like heck over 10 minutes. Helped cool us off. Tues.16th Cooler. Rained last night. Hatch and overhead sweated over my bunk for 2 hours. Caught drip in my helmet. All guns aboard fired today. Testing. Had gas mask drill and more camouflage instruction. Listened to Lt. Col. Roosevelt give lecture. Plenty smart. Wish we knew where we are going. From the diary of Harry Hays, February 1943

Jimmy Roosevelt was the oldest son of President Franklin Roosevelt. He was the Commanding Officer of the Fourth Marine Raiders. Upon arrival in Espiritu Santo, the Marines were supposed to leave the SS President Polk for special operations. However, Roosevelt got sick enough to require an immediate return to the U.S. He later served in the Aleutian Islands Campaign and was awarded the Silver Star. O'Flynn shared that during the ride across the Pacific about half of the Fourth Raiders sold their stiletto knives for gambling money. "I guess the Commander of the Fourth Raiders found out they were selling their stilettos, so they put the word out to all the Navy: You get caught with one of those knives, you get court-martialed," O'Flynn said. "So we had to give them back to them."

Wed 17th Hot again. Had lecture on tropics. How to treat the natives, make spears, what and what not to eat, swimming, etc. Discussed welding. Talked to Master Assist Navigator Master Tech Sgt. $300 a month. Pretty good. Sure getting tired of this. Thurs. 18th Haven't gambled yet. (Oh, yes on Mon. 15th received Shell Back Card on crossing Equator). Ship sighted off port side. First thing we've seen since we've been out. Had lecture on scouting parties, shooting at aircraft, digging in, etc. From the diary of Harry Hays, February 1943

The Crossing the Equator Ceremony is considered a rite of passage for all oceanic Sailors. You are a Pollywog until you have crossed the line and completed the initiation; then you become a Shellback. A mock court is assembled from Shellbacks consisting of King Neptune, Davy Jones, various other members of royalty and a Royal Baby. All Pollywogs go before the court. All are found guilty, and all suffer several punishments. These include the use of truth serum (hot sauce and other items from the galley), crawling on hands and knees across the ship's decks while being sprayed by fire hoses, being swatted with short lengths of firehose and crawling through large containers of mixed, wet garbage. Finally you return to the court and swear your everlasting loyalty by kissing the Royal Baby's grease coated bare belly. Those who have participated in this event never forget it.

O'Flynn remembered the ceremony, saying, "It is a big deal to cross the equator and time zones." He added, "We had one officer on there, that old King Neptune. He was in charge of this organization, so he was just a Commander or something in the program. So we had this Lieutenant, he was a real sharp-looking Italian guy. He had a real sharp mustache, but they told him we have got to damage your mustache, but it is such a good-looking mustache, we will only take off half of it. That was the only real flap we had. The guy was a little upset about losing half his mustache."

After the crossing the line ceremony, all hands, officers and enlisted, received a card to keep in their wallet and an entry was made in each man's service record. Below are the crossing the line cards for Hays and Little, note the same signatures and "CASU Eleven stamp." It is unknown why Hays' card has the name Jimmy Roosevelt written on it, maybe just a souvenir signature, while Little's card has "USS Polk" written over "Vessel X."

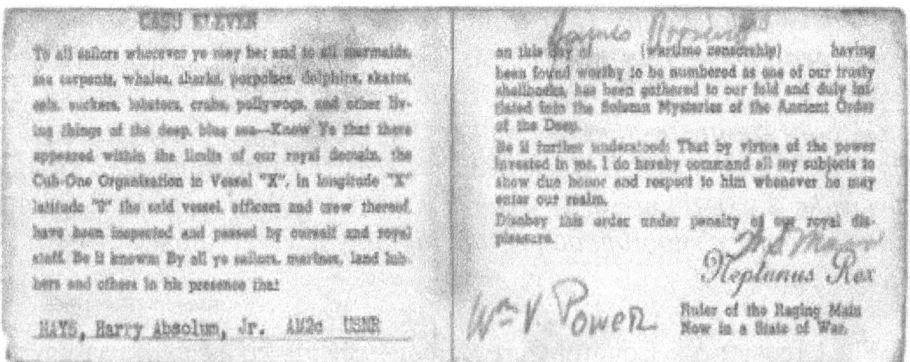

Crossing equator card. From Harry Hays.

18

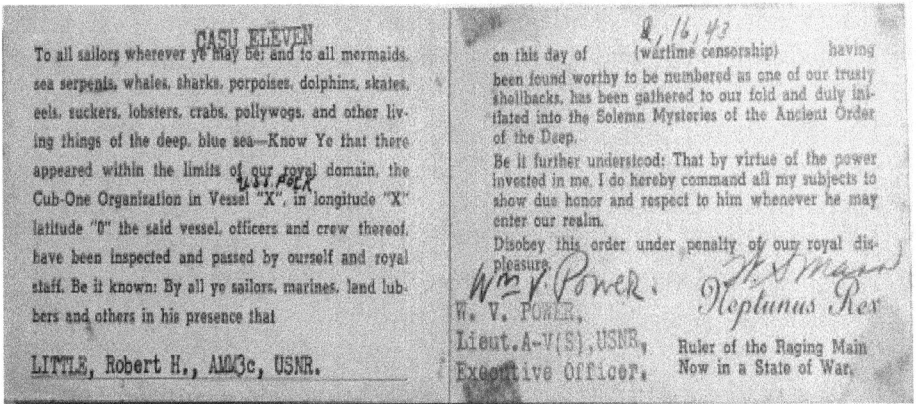

To all sailors wherever ye may be: and to all mermaids, sea serpents, whales, sharks, porpoises, dolphins, skates, eels, tuckers, lobsters, crabs, pollywogs, and other living things of the deep, blue sea—Know Ye that there appeared within the limits of our royal domain, the Cub-One Organization in Vessel "X", in longitude "X" latitude "0" the said vessel, officers and crew thereof, have been inspected and passed by ourself and royal staff. Be it known: By all ye sailors, marines, land lubbers and others in his presence that

LITTLE, Robert H., AMM3c, USNR.

CASU ELEVEN

on this day of   (wartime censorship)   having been found worthy to be numbered as one of our trusty shellbacks, has been gathered to our fold and duly initiated into the Solemn Mysteries of the Ancient Order of the Deep.

Be it further understood: That by virtue of the power invested in me, I do hereby command all my subjects to show due honor and respect to him whenever he may enter our realm.

Disobey this order under penalty of our royal displeasure.

W. V. POWER,
Lieut. A-V(S), USNR,
Executive Officer,

Neptunus Rex
Ruler of the Raging Main
Now in a State of War.

Crossing equator card. From Robert Little.

Friday 19th Bad weather. Rained heavily last night and spurts today. Its 5:00 P.M. and has started to rain cats and dogs. Sat. 20th At about 10:30 P.M. last night we had a scare. The word was passed that a Jap cruiser was chasing us and was being chased by a U. S. Battlewagon and Tin Can. Everyone had to get dressed and stand by for raid emergency. What a night. Hot as heck down below and always a bunch of kids prowling and cutting off the air. I just stayed in my bunk and perspired. Overhead sweated on me again. Got up at 5:00 A.M. Had lecture by Marine who was one of the first to land at Tulagi and Guadalcanal. Plenty tough. Very interesting. I heard it twice. Payday. No dough. Lots of gambling going on. One of the prettiest sunsets I've ever seen. Clouds passing each other in different directions. Rainy and rough. Can see the rain coming 15 minutes before it hits. From the diary of Harry Hays, February 1943.

There was no mention of a Japanese ship in pursuit in the SS President Polk log book. It appears to have been one of the many rumors that made the rounds.

Tulagi is an island approximately 12 miles due north of Guadalcanal that was occupied by the Japanese in early May 1942. The Japanese quickly initiated their plans to build a seaplane base and use the harbor for anchoring their ships. On 7 August 1942, the First Marine Raiders landed and took these facilities after a day of hard fighting. Simultaneously, Marines were landing with little resistance at Guadalcanal. After Tulagi's capture by Naval and Marine forces, the island hosted a fleet of Navy PT boats, including John F. Kennedy's PT-109 and seaplanes used to rescue many downed pilots.

The "No dough" diary entry refers to pay day and money. Sometimes the officers would decide the gambling was taking advantage of the inexperienced and cut off the money flow – the owed pay would just be delayed until later. This strategy did not work, the guys would use buttons or bottle caps and keep a "tab" to be collected later.

Sun. 21st up at 4:00 A. M. Showered and washed clothes. Now 6:15 A. M. Seas calm, but a little rain. Waiting to go to Mass. Went to Communion. Have entire day off. We skip Monday as tomorrow will be Tuesday 23rd. Nothing to do. Sure wish I was home. Had date-line party. Heard 3 Marines entertain. Piano, guitar, harmonica. Swell. From the diary of Harry Hays, February 1943.

The SS President Polk crossed the International Date Line and continued heading west, which meant the crew effectively lost a day - Monday. This event comes with another Navy ceremony for all first-timers, but it was much less of an event than crossing the Equator.

Tuesday 23rd Raining and hot. Another gas mask lecture. Cleaned rifle. Windy. Nothing new. Wed. 24th Nice weather. Lecture on how to abandon ship. I have a cold and cough. Heat, wind, rain etc. Too much change. Had a lecture on what to do with money. Allotments, insurance, etc. From the diary of Harry Hays, February 1943.

The SS President Polk was getting very close to Port 3, Espiritu Santo. Rigsby noted during his interview, "There were two islands there and right out in front of us there was a channel or body of water between them. And then right out in front of that there was another little island. In order to sail in, you have to go around that island. As we were coming in, they were pointing out the fact that right there that was where the President Coolidge was sunk."[6] It was yet another stark reminder of the dangers the men faced when engaged in war.

"On October 26, 1942, at 0930 while approaching eastern entrance to Segond Channel, Espiritu Santo Island, New Hebrides, vessel (President

Coolidge) entered U.S. mine field and struck a mine," according to the SS President Coolidge's deck log. "At 0930 and 30 seconds second mine struck. Port list commenced immediately. Vessel was swung to starboard and beached at about 0935. Abandoning of ship by passengers commenced immediately thereafter. At 1052 vessel turned over on port side and at 1053 slid backwards off coral reef and disappeared beneath the water. One of the merchant marine firemen was killed by the first mine, and it is believed that an army officer lost his life in the accident."[7]

USS President Coolidge abandons ship near Espiritu Santo. Navy Photo.

The President Coolidge was in a "U.S. mine field," and this terrible tragedy could have and should have been averted by clear communications to all arriving vessels. Luckily procedures were clearly delineated and distributed by the time of the SS President Polk's arrival four months later. They avoided the protective minefield that was laid to keep the Japanese out.

Thurs 25th coasting in. All packed. Islands on all sides. Very pretty. Saw airplanes today. Also saw ship going back. Wish I was on it. Destroyers and patrol boats circling us. Beach sure looks good. Can see roads, building, tents etc. Trucks and jeeps on road. 5th & Bdwy. Swell harbor. Full of ships.

One flat top. Planes overhead. Dropped anchor at 3:29 P.M. Had to sleep aboard. No blackout at all. From the diary of Harry Hays, February 1943.

The SS President Polk dropped anchor in port at Espiritu Santo at 1529 and had five shackles of anchor chain in the water to reach the 300 foot bottom and set the anchor. "Anchor chains are usually made up of 15-fathom [90 Feet.] lengths, called 'shots,' which are connected by shackles. In reporting how much chain is in the water the words 'shot' and 'shackle' are synonymous."[8]

It was expected that Espiritu Santo would be immune from air raids or attacks by Japanese ships. This proved to be a false assumption. The Japanese could reach Espiritu Santo with their bombers and ships, they just rarely chose to do so.

Espiritu Santo is located at latitude 15 degrees south and longitude 167 degrees east. The War History of Commander Air Center Espiritu Santo reported some interesting facts about this strategic location. "The geographic location of Espiritu Santo made it a great strategic and tactical value in the early phases of the Pacific War. Lying as it does just north of New Caledonia with its rich mineral deposits, it was the ideal base for the Japanese to use as a spring board for attack on New Caledonia. A base in the New Hebrides could also have been utilized to great advantage as a rear area supply and concentration point for an attack on Australia and could have been built into a strong bastion in the Japanese outer perimeter of defense, especially for a naval base at Segond Channel. From the point of view of its potentiality both as an offensive and defensive position in the South Pacific, the occupation of Espiritu Santo by the Japanese would have represented a serious threat to the success of the Allied Forces. By the same token its value to the Allies in operations against the Japanese was equally potent."[9]

Fortunately the Americans arrived at Espiritu Santo first. "At daybreak in June 1942 the first task force of the American Invasion steamed into the Segond Canal [Channel] at Espiritu Santo, and they commandeered all the available barges and punts, etc., belonging to the French Trading Companies, and Plantation Owners residing in the Segond Canal [Channel]. It was indeed a hurry rush up to land all the troops with their gear, arms, rations, trucks, and machinery etc., before the Japanese got wind of what was happening at the south end of Santo. At that time there were no piers or cranes to assist in the unloading of the punts, and as the rain pelted down on the Americans they just had to keep going to accomplish the job in time."[10]

Five weeks later, in July 1942, U.S. Army bombers were flying the 650 miles from their base at Bomber One on Espiritu Santo to Guadalcanal. The bombers were conducting pre-invasion strikes against the Japanese forces who were building what would become Henderson Field. On 7 August 1942 U.S. Marines landed. By the time the SS President Polk pulled into Segond Channel in early 1943, Pallikulo Bay and Segond Channel had enough pier space to dock about seven Victory cargo ships at one time. And the Marines had ended up a horrific seven months fighting to take Guadalcanal from the Japanese.

"By December 1943, Espiritu Santo had grown into the largest Naval Base, Naval Supply, and Naval Air Base west of Pearl Harbor."[11] Long-time resident of Malo Island, Matt Wells, lived near enough to Espiritu Santo to see ships coming and going. He said that he "counted over one hundred and fifty ships of all kinds lying at anchor ready to load and unload their cargoes. Over two hundred ships a month, merchant and warships were fueled here from the largest tanks in the South Pacific."[12]

Friday 26th Hot with very little breeze. Flies are beginning to appear. Yesterday 5 Natives came alongside in an outrigger. Sang for money. One came aboard and bought a mandolin for $15.00. Sure have big feet. Latest scuttlebutt is that we go ashore tomorrow at 8:00 A.M and wait until the Raiders disembark then reload and go on to Guadalcanal. Will have to wait and see. Raining like heck again. From the diary of Harry Hays, February 1943.

Little also mentioned the flies of Espiritu Santo. "Down in New Hebrides, they had flies that would eat you and pick you up and carry you around. They, ah, yeah, they were a real nuisance. I don't know why they were there. The place was lousy with flies, and then when I went up to Guadalcanal, there wasn't any flies to amount to anything. Very, very, few flies."

In the original plan the SS President Polk was to stop in Espiritu Santo and quickly drop off the Fourth Marine Raiders and their gear. This should have taken two or three days. Then the ship was to proceed to Guadalcanal. However, this did not happen. Since the ship was so poorly loaded in San Diego, the men experienced a 17-day delay.

The U.S. Pacific Fleet, Fleet Air, West Coast in their war diary used the example of CASU-11 and the Fourth Marine Raider Battalion as testimony of the negative impact poor loading had on forward operations. It was expected the SS President Polk would rapidly off-load the Marines in Espiritu Santo, where a modified destroyer was waiting to take them to conduct their secret missions. The SS President Polk would then get underway with CASU-11, delivering them to Henderson Field, Guadalcanal. The Fleet Air report stated, "CASU-11 and a Raider Battalion were loaded at San Diego for Espiritu Santo, and the equipment of the two units was mixed in every hold of the ship, placing the entire cargo on the ship's lighters [small vessels belonging to the SS President Polk for transporting offloaded gear from ship to shore], thereby tying up the lighters, which required sorting the two unit's equipment and reloading the CASU. This process took a week and used a large number of men."[13]

There was a complete lack of understanding in San Diego of "unit or combat loading," which is the separate loading of CASU and Marine equipment. Separate loading would have enabled the expeditious offloading of the Marines in Espiritu Santo, and the SS President Polk would have been underway for Guadalcanal in four to five days instead of 17. In addition, filling all the SS President Polk's lighters also hindered the rapid sorting of cargo. This disorganization surely led to heated conversations on the pier between the Marines and Sailors as they separated and reloaded heavy crates in the hot tropical sun and humid air. Fourth Marine Raiders Daniel Marsh stated, "I have nothing to say about the unloading operations of the President Polk. Such activity would be hazardous to my health and well-being. Eventually the holds were unloaded and the Raiders were once again busily involved in setting up camp."[14]

Merillat, in his book *Guadalcanal Remembered,* defines combat loading this way, "Combat loading calls for stowing rations, equipment, weapons, ammunition, and other material in such a way that things most urgently needed in combat will come off first, in amounts required by the units aboard a particular ship. It calls for specialized and exacting skills, developed by the

Marine Corps and Navy over years of planning and exercises."[15] The stevedores of San Diego's waterfront were going to gain an understanding of combat loading in short order.

Sat. 27th Finally got ashore around 3:00 PM What a day. First we ate coconuts right off the tree. Also limes. No more coconuts for me. Were advised they start dysentery. Hiked about 10 miles to our camp. No one knew where it was not even the officers, but we found it. Last 2 miles in a driving rain. Mud up to our knees. Had left-over chow. But better than aboard ship. Black as the ace of spades. Seabags and packs all piled in a heap. What a mess! Had to find tent to sleep in. Fell in Fox Hole with full pack and rifle. Had to walk back a mile to get mosquito net. Finally to bed. From the diary of Harry Hays, February 1943.

A fox hole was a place to hide when bombs are dropping. It was usually dug by the members of the nearest tent and large enough to hold the four members of that tent. They were also dug near all work sites. Everybody participated in the digging, and this work was generally a first priority. Hays was lucky he did not break an arm or leg when he tumbled into one!

Fox Hole. Marine Corps Photo.

25

Sun. 28th Had chow. Very good. Coffee swell. Tried to wash clothes in ocean. No go. Went for a swim. Hung clothes to dry. Then got word we had to move to other side of island. How's that for efficiency? Hot. Sweat rolling off us. Finally moved. Much better place. From the diary of Harry Hays, February 1943.

At the time of CASU-11's arrival at Espiritu Santo, work was ongoing at a new, nearby bomber airfield. While bulldozing, an unexploded Japanese bomb was discovered. This required all bulldozer work to end - only shovels could be used. Volunteers were requested. One of CASU-11's Sailors who volunteered was Aviation Machinist Mate Third Class Ferdinand J. Isenberg, who, upon successful completion, received the following note signed by the Commanding Officer, Lt. Cmdr. Schlossbach and placed in his service record:

"2-28-43 Volunteered for the dangerous detail of digging for a Japanese bomb which landed near Bomber Strip, Espiritu Santo when volunteers were asked for by Cmdr. CURREY, Commanding Cub ONE, Receiving Station, Espiritu Santo. They worked 8 days on this detail; were complemented for their work by Ensign J.S. WHEELER, demolition officer in charge."[16]

The staff of Adm. Nimitz, Commander in Chief Pacific Fleet (CINCPAC) produced monthly *Operations in Pacific Ocean Areas* reports that quickly summarized the Navy's actions for the previous month. The following words were from the February 1943 report.

The first major event of February was the 'total and complete defeat of Japanese forces on Guadalcanal effected 1625 on 9 February,' as reported by General Patch. This came quickly on the heels of an amazingly successful and surprising Japanese evacuation of 11,706 men from Guadalcanal over three nights, the fourth through the seventh of February. The second was our occupation on 21 February of the Russell Islands preparatory to our further advance in the Solomons.

Adm. Nimitz stated, 'Our victory on Guadalcanal was accomplished against the opposition of the best of the enemy's armed forces. In the air the Japanese threw into combat their best carrier and land-based squadrons and suffered severely in destruction of planes and experienced crews.[17]

These monthly CINCPAC reports will be given at the beginning of each month's diary entries to provide a top level explanation of major war events that occurred in the Southwest Pacific.

Even though the Marines now controlled all of Guadalcanal, the Japanese were not finished. Instead of trying to defend Guadalcanal with ground forces, the Japanese were now going to attack the Allies using Japanese aviators and aircraft from island airfields north of Guadalcanal.

March 1943

CINCPAC *Operations in Pacific Ocean Areas* report for March 1943: In March we continued improvement of our bases in the lower Solomons preparatory to future moves towards Rabaul. The principal development was in the Russell Islands, occupied during the last week of February. The fighter strip, under construction on the east end of Banika Island, progressed rapidly and was almost ready for use on 31 March. Enemy strikes during March against (our positions on) the Russells, Tulagi, and Guadalcanal, caused minor damage without real hindrance to development of these bases.

While most flight operations during March were defensive in nature and in response to inbound Japanese aircraft, there were some offensive activity. These were devoted to bombing air bases in the Mid-Solomons, especially at Vila and Munda Fields. During the month the air forces in our Southwest and South Pacific commands dropped about 425 tons demolition and incendiary bombs from Munda Field to Rabaul, causing some destruction of supplies and installations.

Mon 1st Shower. Ahhh! Chow good. Plenty of flies. Mosquitos not bad you must stay in after dark. Went swimming. Not very good. Tues. 2nd Washed clothes. Went swimming in another spot. Much better. Received first mail from my Baby and the Folks. Answered both. From the diary of Harry Hays, March 1943.

Letters from home were an important element in maintaining morale. Because Espiritu Santo received increasing numbers of daily cargo shipments to support the growing offensive up the island chain, CASU-11 received mail almost daily. Once CASU-11 arrived at Henderson Field, Guadalcanal, they continued to receive mail frequently. Letters would be saved and read multiple times, and often times passed to friends to also read. Sometimes letters would be read out loud. Little's mom was concerned about her son and his whereabouts, so she asked in a letter, "Where are you?" Little, in answering this letter, noted that he was not allowed to tell her his location. However, later, in this same letter, Little asked a question, seemingly about a neighbor. He wrote, "How is Mr. Henry doing on his 'canal' connecting Deep Creek and Mr. Wilson's pond?" Not very subtle, but his subterfuge got past the censor, and the Little family knew where their second son was located.

"We wrote letters; you didn't even have to stamp them - put free up on the corner," O'Flynn said. "I never wrote too much. Two or three times a month, I'd write a letter home. I didn't have time. As a matter of fact … this old John Parker left about a month before we did, and I gave him a letter to mail when he got back to the States. I didn't write again until I went home."

Wed. 3rd Washed Seabag and Hammock. A little sunburned, especially my head. Shaved and showered. Raining now, 2:00 P.M. Nothing new. Thurs. 4th Our Lt said to be ready to move in 4 or 5 days. Nuts! Went swimming then showered in driving rain. Took advantage of rain and washed clothes. Fri. 5th Were told that 50 of us had to move to Hospital for duty. Lashed all up and then secured again. Another bad move. They can't expect us to have much confidence in them with such inefficiency. From the diary of Harry Hays, March 1943.

Rapidly changing orders can be very frustrating. The order to "lash" up, or pack up, for a few days of duty at the hospital, and then, after everyone is ready to hike, being told that the order is cancelled is just one example. Paul Fussell, in his book *Wartime,* described World War II military leaders, "For most, this is the first war they've fought in. They are neophytes and amateurs, plucked from civilian life to engage in deadly on-the-job training in an unfamiliar atmosphere of rigid hierarchical 'authority,' where orders (sometimes euphemized to 'directives' to spare civilian sensibilities) are not to be questioned and are seldom discussed, no matter how absurd, unreasonable, or based on patently erroneous assumptions."[18] Most officers tried to do their best. However, those tasked with carrying out the orders rarely had the chance to discuss and determine the best course of action with their senior leaders. Sailors Hays, Little, Wilson, O'Flynn, West, Parker and Leal will experience much more of this before they get home from the war.

Sat. 6th Played chess and cards for the first time. Hearts. Walked 4 miles to canteen. Threw at black bat in tree. Sun. 7th Went to show. "Random Harvest." Planned to go to church. Poured rain. That's all. Mon. 8th My head is peeling from sunburn. Had to move back to ship. Gear all dirty again. Loaded lumber until 8:00 P.M. No fresh water. Another mess. Tues. 9th Have been put on watch detail. Supernumerary. 4 on 8 off. Took shower and shaved. Lucky to find fresh water. Was sick yesterday. Lost my dinner. Atabrine, I think. From the diary of Harry Hays, March 1943.

Atabrine is a malaria medicine, typically prescribed once a day with a meal. This was just the beginning of CASU-11's introduction to mosquitos and malaria.

Wed. 10th Officers couldn't get their working parties straight as they mustered everyone at 12:45 last night. Another mistake. Have good bunk now. Hope I get to keep it. Thurs. 11th Lost my bunk. Fri. 12th Made P.O. of the watch. 4 on 12 off. Have 12 to 4 again tonight. Still no mail. From the diary of Harry Hays, March 1943.

P.O. is Petty Officer, and, in this case, it is the person in charge of a work group or security team.

Sat. 13th Nothing new. Sun. 14th Couldn't go to church. Had 8 to 12. Got a fair bunk. Because we weren't arranged right when we came aboard, practically everyone had to move. Was until 9:30 P.M. finding out where my 12 to 4 watch slept. Got 1 1/2 hours sleep. Mon. 15th Pay Day $83.00. Not drawing. Shoving off at 4:35 P.M. Scuttlebutt is Guadalcanal. Am praying we make it. From the diary of Harry Hays, March 1943.

As of 15 March 1943, Hays had been on the SS President Polk for the previous seven days. Despite the unloading and reloading of the SS President

Polk, there was considerable confusion about where men would sleep. His comment about not drawing meant he decided to leave his pay on the Supply Officer's paybook for a later withdrawal.

Part Three - Guadalcanal, Solomon Islands

The gossip for mid-March 1943 on Espiritu Santo was about Guadalcanal. Even though the Marines had captured Guadalcanal in February and declared it free from all enemy ground forces, Espiritu Santo was still completely focused on supporting operations on Guadalcanal. Ships and airplanes were headed to and from Guadalcanal daily. The men of CASU-11 certainly heard the rumors and knew they were headed to Guadalcanal, since no one was suggesting that CASU-11 would stay on Espiritu Santo. The men of CASU-11 were listening to every bit of scuttlebutt, dissecting every sentence and hearing things they did not want to hear.

There were three problems between CASU-11and Guadalcanal: Japanese bomber aircraft, Japanese destroyers and Japanese submarines. The submarines were located in an area called "Torpedo Junction." Rigsby remembered this danger years later and shared it during his interview, "I know we went through one area. You know the scuttlebutt was, oh, this is called torpedo junction and you were lucky if you get through this area. They had a couple of destroyers now following us. Don't know if they would have been able to do much if there was a Jap sub there who was bent on sinking us. It would have been a rich problem because they had a couple of thousand men aboard that would be a difficult replacement problem."[1]

John Prados in his 4 March 2013 blog *Torpedo Junction* further explained, to stay on Guadalcanal, "Allied invaders needed seaborne supply from their nearest bases: the New Hebrides island of Espiritu Santo, about 400 miles southeast of Guadalcanal, and Noumea, another 450 miles south.

From Espiritu Santo, a fast ship could reach the island in a day and a half. A merchant vessel on the Noumea - Guadalcanal run needed almost four days - and a naval escort."[2] To stop this first Allied offensive in the South Pacific, Japan needed to cut these shipping lanes. While "the resulting aerial and ship battles are famous, the underwater offensive the Japanese waged in the Guadalcanal supply corridor"[3] is less well known. However, it was talked about enough to earn the moniker of Torpedo Junction. The men of CASU-11 had a right to be scared - Japanese submarines were expected to patrol the waters they were going to traverse over the next few days.

More specifically, Torpedo Junction was a broad area around a line that extended east-west between Ndeni in the Santa Cruz Islands to San Cristobal Island. The shipping routes from Noumea and Espiritu Santo to Guadalcanal and the great circle route from Hawaii to Australia crossed through this area. Prados closed his blog with, "Careful analysis shows Japanese submariners to have been as effective there as the empire's surface navy - that is, until Japan's submarine campaign foundered on the empire's rigid naval doctrine. The Imperial Navy suddenly demoted I-boat submarines to supply duty - a step with no precedent, taken in desperation to succor starving Japanese soldiers on Guadalcanal. This supply order effectively ended the Torpedo Junction blockade."[4] And even though this submarine pull-out occurred a few months before CASU-11 ran these waters, no one on the Allied side had any idea this had occurred. So the SS President Polk had four escorts for the crossing of this dangerous area. As it turned out, this attempt by Japan to deliver food by submarine proved to be less effective than if they had stayed on duty in "Torpedo Junction," sinking Allied shipping. CASU-11 was lucky.

The SS President Polk deck log reported at 1705 on 15 March 1943 "underway and stood out from Espiritu Santo. 1736 Departed Port 3." The ship was part of Task Unit 34.4.8 for the transit to Guadalcanal. Ships in company included another troop transport, the SS President Tyler. Also in the task unit were four escorts, the USS Sands (APD 13), a Destroyer High Speed Transport; the USS Ellet (DD 398), a Destroyer; the USS Morris (DD 417), a Destroyer; and the USS Breese (DM 18) a Destroyer Light Minelayer.

Tues. 16th Pres. Tyler with us. 3 Destroyers for convoy. 2 alerts today. Slept in Clothes, etc. Had 8 to 12. Am sick. Taken off watch list. One destroyer supposed to have dropped 2 charges last night. Wed. 17th Feel better. Taking sulfa pills. It is 1:00 P.M. and we just had a real alert. Boy those tin cans really move. All secure now. From the diary of Harry Hays, March 1943.

The unceasing zig-zagging of every ship was a constant reminder that Japanese submarines could be nearby. While Hays doesn't reference the emotion associated with the alerts on 16 and 17 March, these alerts marked the first time the men of CASU-11 thought they were in real danger. The Ellet dropped a depth charge on 16 March in an attempt to sink a suspected Japanese submarine. The men's anxiety, excitement and fear likely grew as the SS President Polk sailed closer to Guadalcanal.

The following are the deck logs of the ships in Task Unit 34.4.8, as they told their version of the story on sailing to Guadalcanal:

15 March 1943
Ellet[5] - 1740 Underway - Senior Officer Present Afloat and Commander 32.4.8 in SS President Polk. [This means the senior officer for all of these ships was riding the SS President Polk.]
Breese[6] - 1830 Took departure for Guadalcanal, on course 000(T), speed 13 knots; zigzagging according to plan No. 17.

16 March 1943
Morris[7] - 0120 Ellet attacked underwater contact.
Ellet - 0120 Dropped one (1) 600-lb depth charge on fair sound contact. Continued search until 0350 then proceeded to join task force. Search results negative. [Depth charges are bombs dropped in the water near sonar targets in hopes of sinking a submarine.]
Morris - 2230 Sighted eight unidentified ships bearing 330 (T).
Morris - 2235 Ships identified as friendly.

17 March 1943
Ellet - 1009 SC radar picked up aircraft contact bearing 310 (T) distance 31 miles. Aircraft was sighted and identified as Japanese twin-engine land plane at 19 miles. Plane circled formation and departed to southeastward.

Breese - 1250 Ellet established possible submarine contact by echo-ranging bearing 270 (T).

Breese - 1259 Ellet contact established negative.

18 March 1943

Ellet - 1950 Went to general quarters - unidentified planes picked up on radar bearing 300 (T) distance 50 miles.

Ellet - 1951 Radio Cactus [Guadalcanal's codename was Cactus] reported enemy planes approaching.

Ellet - 2018 Enemy planes dropped flares bearing 200 (T) from Koli Point. No bombs were observed to drop.

Thurs. 18th coming into Guadalcanal. Looks the same as Santos. Beautiful sunset. Fri. 19th Guadalcanal on one side Tulagi on the other. Saw 6 foot shark today. It is 4:00 P.M. Task Force coming in. 4 Cans 1 Heavy Cruiser. From the diary of Harry Hays, March 1943.

"Cans" or "tin cans" are destroyers - to those who rode in them it seemed they were very thin skinned. They actually had a steel hull, it just felt like the hull was made of tin, especially to those who had seen Japanese bullets go through it.

Part Three A - The Importance of Guadalcanal and CASU-11

The Sailors of CASU-11 did not know for sure where they had been headed or what their mission entailed. However, Admiral Nimitz, Commander Pacific Fleet, had a definite plan that would stop the Japanese and place the United States military on the offensive. The black line on the map below outlines the area controlled by Japanese military forces by early 1943. Back in the United States, senior officers on the Joint U.S. and British military staff headquarters were becoming extremely concerned with the size of this area and the possibility of Japan cutting the shipping lanes between Australia and the United States.

The threat to Australia was very real. By early 1942, the Japanese military had seized Rabaul, New Britain, from Australia and turned it into a fortress with numerous anti-aircraft batteries and 100,000 Japanese troops. From this location Japanese forces had successfully launched attacks against Australia and deployed forces to establish airfields and bases along the Solomon Islands. In May 1942, three Japanese midget submarines entered Sydney Harbor and attacked shipping. Two of the three subs were attacked by the Australians, and the Japanese crews committed suicide by intentionally letting their boats sink. The third sub successfully sank a local ferry and 21 Allied Sailors on board were killed. On 8 June 1942, two submarines shelled the eastern suburbs of Sydney and the city of Newcastle with no reported injuries or deaths. In May 1942 the Japanese invaded Guadalcanal in the Solomon Islands. August 1942 photographs from reconnaissance aircraft showed the Japanese already had a long-range seaplane base operating at Tulagi, twelve miles north of Guadalcanal, and the Japanese were actively building a runway on Guadalcanal. This airfield would allow the Japanese to execute their planned Fiji and Samoa operation, dubbed FS Operation. The successful completion of the FS Operation by the Japanese would give them control of the sea lanes and enable the invasion of Australia.

West Pacific map from Keese and Sidwell.[8]

Nimitz knew that the take-down of Rabaul was necessary before commencing the campaign towards the Japanese mainland. His question was: Where is the best location to position Allied forces to take on this task?

While the United States had a presence at numerous places south of the shipping lanes, Nimitz knew he needed to establish a position north of the shipping lanes to provide complete lane protection. Additionally, due to fighter and bomber airplanes' 300-mile operational range limits, he knew he would need multiple islands to provide stepping stones to reach Rabaul.

The first stepping stone island chosen was Guadalcanal, and it was located north of the sea lanes connecting Australia with the United States. It was considered the most remote, yet populated island in the world. The largest of the Solomon Islands at 2,000 square miles, Guadalcanal is volcanic in origin.

The southwestern half of the island is mountainous with peaks reaching 8,028 feet and dense forests. The northeastern half of the island was largely plains with numerous coconut plantations.

Admiral Nimitz's plan was to attack Guadalcanal and begin pushing the Japanese north to Japan. With a shortage of aircraft carriers, he needed land-based airfields manned with personnel specifically trained to support and repair naval aircraft. During early 1942, as the war in the Pacific had begun to unfold, U.S. Navy leadership had recognized they were going to need more airfields for launching airplanes against Japanese forces. As Navy planners studied maps of the western Pacific, they discovered thousands of islands many had never heard of before. When Navy intelligence officers started marking many of these islands with Japanese airports and runways, some planners began realizing that the path to Japan was going to require the use of these airfields. This tactic became known as "island hopping." As it unfolded, the plan consisted of U.S. Marines performing amphibious landings on selected islands, followed by Naval Construction Battalions, or Seabees, repairing and improving the captured runways and/or building new ones. Then, Marine, Navy and Army Air Corp aircraft arrived using the newly captured airport to assist the Marines with their next island target.

The next map shows the area of the Southwest Pacific that includes Guadalcanal and Rabaul. Guadalcanal can be found at the southeastern end of the Solomon Islands. Rabaul is a large harbor located on the northeastern tip of New Britain, and it is part of the Bismarck Archipelago. Because of its shape, the waterway lying between the Solomon Islands was known as "The Slot". It was frequently used by the enemy for travel between Rabaul and Guadalcanal.

Guadalcanal was the first island attacked by the United States Marines in the Pacific campaign. Admiral Nimitz had begun the offensive that would ultimately bring the war to its conclusion in 1945. Co-incidentally the Allied transition from taking and defending Guadalcanal to launching offensive warfare up the Solomon Island chain began on the same day, 9 Feb 1943, the men of CASU-11 departed from San Diego.

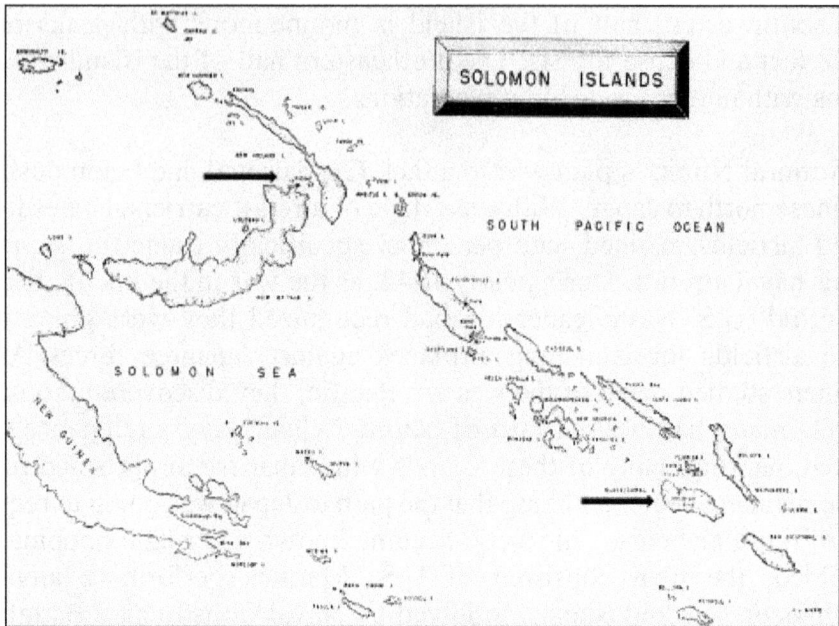

Chart of Area Surrounding Guadalcanal (Lower arrow) and Rabaul (Upper arrow). Marine Map.[9]

In his book, *Islands of Destiny* John Prados declared,

By early August [1942] the Japanese had almost completed the Guadalcanal airstrip. Then, one day at dawn, the waters offshore were filled with Allied vessels. Cruisers and destroyers bombarded Japanese positions on both Guadalcanal and Tulagi. Transports began to lower landing craft and fill them with troops."[10] [The march to Japan had finally begun, and the landing was called Operation Watchtower. The next seven months would be an unrelenting test under fire as the U.S. Marine Corps met the Japanese forces for their first lengthy battle.]

While the Marines were chasing the Japanese through the jungle, other Marines commenced completing the runway by hardening the runway surface with crushed coral and gravel and extending it in preparation for heavier US aircraft. Marines filled bomb craters and continued grading, working from the ends towards the middle. Near the middle of the runway a huge hole needed some 5,000 cubic feet of dirt to fill it and the Marines' bulldozers were still offshore loaded on the ships. In a monumental effort, mostly by shovel, the Marines managed to fill the hole and open the airfield on 12 August. Two days before this opening General Vandegrift named the field after Major Lofton R. Henderson, a Marine Squadron commander lost during action in the Battle of Midway.[11]

The 6th Naval Construction Battalion[12] arrived 1 September to continue the work. They found a 3,800 foot runway made of unstable muck filled with ruts from numerous landings combined with very rainy weather. They commenced grading the runway with a crown to aid with drainage. Then on 25 September they started laying interlocking metal Marsden matting and extending the runway an additional 1,300 feet. This work was completed October 1942, and at the end of November Henderson Field was designated as "Bomber One" with a well-drained main runway consisting of 3,400 feet of metal matting continuing with 1,600 feet of gravel, crushed coral and clay.

Seabees installing interlocking Marsden Matting at Henderson Field.
Navy Photo.

Marsden mats were steel, perforated, panels measuring ten feet long by 15 inches wide with a thickness of ¼ inch. Each panel weighed 70 pounds. Marsden mats could be assembled into a runway very quickly after a solid, well compacted and elevated subgrade was prepared. It greatly reduced both dust and mud in the landing area and aircraft parking spots. Most of the time that CASU-11 was on Guadalcanal, the men worked on this metal surface. It could be a little slippery when wet but otherwise was much better than working in the mud. For an aircraft runway, Marsden mats beat mud and grass runways which became rutted in wet weather. After bomb damage, a crowbar could be used to remove the bent and broken mats, then coral was compacted into the hole, and finally, new mats were locked into place. On the negative side a hard landing could bounce a small fighter 30 feet, and heavy bombers occasionally made the mat move and bunch up into waves. This required a bulldozer to pull the mats back into place.

Simultaneously with the effort on the main runway, three flight strips were being constructed to supplement Henderson Field. By early October 1942, the engineers had finished a fighter strip, called flight strip number one, about one mile east of Henderson Field. It was a rolled-turf strip, 3,400 feet by 300 feet, and it was constructed in three days. The sage grass was cut to a height of about 18 inches, hummocks leveled, old foxholes filled, and the field rolled. All the equipment was captured from the Japanese during the Marine landing. This field was rolled regularly and it was excessively muddy after a rain. Nevertheless, the use of two fields allowed separation of dive bomber and fighter squadrons and provided further aircraft dispersal. Aircraft parked close together made a better target than aircraft spaced out. Once in October, this fighter strip served all air traffic when "Pistol Pete's"[13] gunfire made Henderson Field unusable. Flight strip number two required grading only. This was accomplished by a single carryall truck and one bulldozer which pulled some Japanese steel trusses rigged into a drag. Flight strip number three was a rolled-turf strip used only for aircraft dispersal and later called the Emergency Field. By the end of October 1942, all three of these temporary runways were in use.

From the beginning, plans for the development of Guadalcanal into a major air base called for the construction of four airfields in addition to Henderson Field and its supplemental strips. On Lunga Point, where Henderson Field was located, two fighter fields were projected. One field was at Kukum and the other at Lunga. About 8 miles to the east on Koli Point, two bomber fields were planned. CASU-11 would work at both Henderson and Lunga Fields.

In November 1942 the 6th Naval Construction Battalion began work on two fighter runways at Kukum Field (Fighter 2), near Lunga Point about a mile west of Henderson Field. Due to heavy evacuations of personnel, representing nearly half its strength, the battalion was relieved on 1 December 1942 by the First Marine Aviation Engineers. At that time, approximately 30 percent of the subgrading for the two runways had been completed. On 1 January 1943, the first strip, which was coral-surfaced, began operations. Between June and July 1943, the 46[14]th Naval Construction Battalion[14] and the 61st Naval Construction Battalion[15] built the second coral-surfaced runway. It was 4,000 feet long by 150 feet wide with 75 foot wide shoulders. It had coral taxiways 80 feet wide, and 121 hardstands for aircraft parking and protection.

On 27 December 1942, the 18th Naval Construction Battalion[16] started construction on Lunga Field (Fighter 1). They had arrived on Guadalcanal two weeks earlier. Their task was completing the runway construction with all possible speed. However, due to bad weather, almost continuous daily Japanese aircraft bombing, and naval bombardment at night, the field was not ready until 9 February 1943. By this date, the runway was 4,000 feet long and 150 feet wide with Marsden steel matting on a gravel base and one taxiway. Two squadrons immediately began operating, and by 7 March 1943, four squadrons were in operation. An additional taxiway and 110 hardstands for light bombers were later constructed at the field by the 46th and 61st Naval Construction Battalions.

On 5 December 1942, the 14th Naval Construction Battalion[17], which had landed at Koli Point early in November after a small detachment of Marine raiders, began work on an emergency fighter runway at Carney Field. The strip was completed in two weeks. On 23 December 1942, construction was started on the main bomber runway. It was 6,500 feet long and 150 feet wide, with a steel mat surface. On 3 February 1943, the construction force was augmented by the 2nd Marine Aviation Engineers, who had arrived at Guadalcanal on 30 January 1943. Heavy rain and daily bombing alerts hampered the construction work, but on 21 March an Army night-fighter squadron began using the field followed by a heavy bombardment unit a few weeks later. Finally, on 1 April 1943 the Koli bomber field, also known as Bomber 2, was completed six miles east of Henderson Field. This field was named after Captain J. V. Carney, USNR Civil Engineer and C.O. 14th Naval Construction Battalion who was killed 16 December 1942 while flying on board an SBD Dauntless.

On 22 May 1943, work was started on Koli Field (Bomber 3) by the 61st Naval Construction Battalion and the runway for heavy bombers was completed in October 1943. The strip located about eight miles east of Henderson Field, 7,000 feet by 500 feet, was built of river gravel with a Marsden mat surface on a silty soil base. Under heavy bomber loads, many failures of the mat occurred, traceable to the poor nature of the sub-soil. Emergency repairs were made to permit uninterrupted operation of planes, by filling coral under the mat to provide adequate support. A new site, with a heavy blanket of stabilized soil, was chosen 300 yards northwest of the original strip, and eventually the original strip was reconstructed. The project was under the direction of the 61st Naval Construction Battalion, with additional equipment and men from ten other Army and Navy commands. The 2nd Marine Aviation Engineers built the south taxiway system that connected Koli Field with Carney Field, and 101 heavy bomber hardstands were provided.

By the time Lt. Cmdr. Schlossbach reached the end of his second month as the Commanding Officer of CASU-11, in May 1943, there were five airports with multiple runways organized and managed as follows:

"Henderson Field (Bomber 1) and Lunga Field (Fighter 1)
Lt. Cmdr. Schlossbach - Airfield Commander and CASU-11 Commanding Officer. Men from CASU-11 worked at both Henderson and Lunga Fields.

Kukum Field (Fighter 2)
Maj. Taylor - Airfield Commander and Commanding Officer 38th Service Squadron.

Carney Field (Bomber 2) and Koli Field (Bomber 3)
Lt. Col. Tidwell - Airfield Commander and Commanding Officer 29th Army Air Service Group."[18]

Eventually the numbered fighter and bomber field names were dropped and the five Guadalcanal airfield complexes became officially known as Henderson Field, Lunga Field, Kukum Field, Carney Field, and Koli Field.

The map below shows the multiple airfields located on the north central coastal plain of Guadalcanal. Between Lunga Point and Koli Point was the bay, beaches and airfields that would be CASU-11's primary operating area. O'Flynn said he never went much more than a mile from this piece of coast line while stationed there. Note the five airfields from left to right, Kukum, Henderson, Lunga, Carney and Koli.

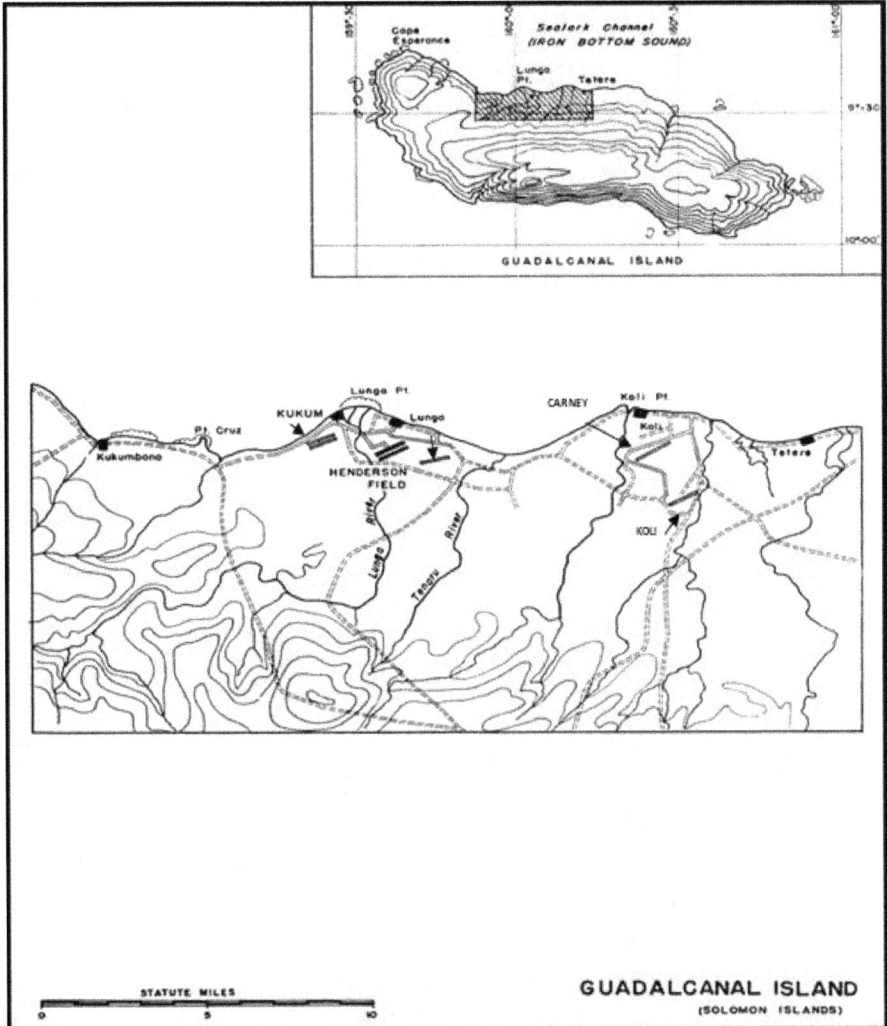

Guadalcanal Airfields. Navy Graphic.

Henderson Field center, CASU-11 camp at arrow. Navy Photo.

Kokum Field at top, Henderson Field center, Lunga Field bottom. Army Photo.

On 1 September 1942 in San Diego, California, the U.S. Naval Advanced Base Unit ACORN (Red) One was commissioned and headed to Guadalcanal. This ACORN would become the partner of CASU-11 on Guadalcanal.

The Navy *Manual of Advanced Base Development and Maintenance* (OPNAV 30-11-A1 Revised April 1945) defined an ACORN and CASU as follows:

An ACORN includes sufficient trained personnel to maintain the runway and aviation facilities in operating condition; to service casual planes; to operate the control tower, field lighting, aerological unit, communication, supply, disbursing, ordnance, transportation and medical facilities; and to maintain berthing and messing facilities to be used by the CASU and aircraft squadron crews when they report aboard. The ACORN generally has a Construction Battalion attached to it to build the airfield and necessary structures.

The CASU includes all the trained personnel necessary to support the maintenance and flight operations of a shore-based Carrier Air Group or its equivalent (60 to 80 aircraft). A CASU is a personnel unit. It has no equipment but depends on the equipment and facilities of an ACORN, consequently ACORNs and CASUs are almost always paired together.[19]

The "Mobile Base" concept described in Ballantine's book, *U.S. Naval Logistics in the Second World War,* was required for this war in the Pacific with the ACORN- CASU aircraft service team being the most applicable components of this logistical construct. In other words, many more ACORNs and CASUs were required than ever expected. It was easier to assemble and train an ACORN-CASU team then to build an aircraft carrier in 1942/1943. Later, when America's shipyards were operating at their maximum capacity, their aircraft carrier delivery rate was an overwhelming factor in the Japanese defeat. In the interim, island airfields and ACORN-CASU teams would have to carry a significant part of the wartime load in the Pacific.

Again, "the CASU and ACORN together constitute a team. The CASU services the aircraft and the ACORN provides and maintains the base and its facilities as well as the equipment and tools needed by the CASU. Close cooperation between the two is essential. After the initial stages of base construction were completed, the ACORN was frequently, but not always, decommissioned and absorbed into the base. The CASU, however, was a permanently commissioned unit, and remained in commission throughout its period of service as a functioning Naval activity."[20] This is going to be the case of ACORN (RED) One which was be decommissioned and absorbed into the Naval Air Center Guadalcanal shortly after CASU-11 arrived.

Usually, Navy aircraft are serviced on aircraft carriers. These ships have the needed parts, fuel, ammunition and maintenance personnel available. During the first few years of the war, the U.S. Navy had a shortage of aircraft carriers. As the Allies captured island-based runways during the Solomons Campaign, the CASUs provided the necessary personnel to maintain their aircraft. This required a quick turnaround of the airplanes, as the island locations were often inside an active war zone. To support this rapid transfer, the CASU was organized as a group of maintenance personnel with minimal equipment or tools, a personnel construct ready and easy to move.

However, the plan of ACORN first, and CASU later or both arriving together was not always successful. From the 1 September 1944 issue of *Naval Aviation News*, "CASU's on advanced Pacific islands run into the same kind of rugged living conditions as frontline troops. Long hours of night work repairing carrier-type planes, sleepless nights filled with bombings, cold food, disease, casualties, foxholes - all of these are a daily chapter in the kind of warfare they are running into. Many units landed right behind the Marines and had to build their bases from scratch before they could start maintaining planes. Their problems are much different from those confronting CASUs operating on the continent. Squadrons had a habit of dropping in before the official reception committee was all set, supplies were delayed and replacement parts lost, but the CASUs kept the planes flying, helping to push the Japs back toward Tokyo."[21]

Bergerud stated in his book, *Fire in the Sky*, that "this [Island airfield] was one of the most primitive places on earth. However, airpower required some of the most sophisticated machines and support available to the combatant nations. Consequently, everything required to fight had to be brought in from the outside. The aircraft themselves, all of the personnel, almost all food, all vehicles, all weapons, all communications, and all medical supplies had to be transported into the theater."[22] Then it had to be maintained and repaired locally. "Both sides routinely pushed aircraft far beyond the hours recommended for overhaul, but regardless of military necessity, there came a time when the plane had to be rebuilt from the ground up. This meant the installation of a completely reconditioned or, better yet, new engine. Major components, such as a wing or landing gear assembly, might need replacement."[23]

From the same September 1944 *Naval Aviation News*, a typical CASU was organized into eight divisions.

ENGINEERING (First) division included personnel to operate everything from the machine shop to the photo and propeller shops.

FLIGHT (Second) division personnel comprised of line engineering crews, parachute loft, and fueling crews.

ORDNANCE (Third) division rearmed aircraft guns, loaded bombs and torpedoes, and kept the plane's armament operating.

COMMUNICATIONS (Fourth) division installed, checked, and repaired all radio gear and handled secret and confidential mail.

Under the FIRST LIEUTENANT (Fifth) came berthing and messing, maintenance of buildings and grounds, the carpenter shop, fire and air raid security, damage control, and maintenance of a garage for care of CASU vehicles.

SUPPLY AND DISBURSING (Sixth) was responsible for procurement, custody, issuance and accounting for all materials required for the operations of the CASU. It inventoried incoming and outgoing aircraft, maintained storerooms for handling material needs of squadrons and the CASU, and maintained office records and handled disbursing.

PERSONNEL (Seventh) and MEDICAL (Eighth) divisions comprised the remainder of the unit's extensive plan of organization.[24]

A CASU's manning numbers varied "depending on whether they maintained 45 planes, 90 planes, 180 planes, 270, or the big 360-plane outfits. They had a standard complement of officers ranging from 16 to 89 and enlisted men from 185 to about 1,500."[25] The larger CASUs were assigned to Hawaii, the West Coast or the East Coast. During the period March 1943 through June 1944, CASU-11 had about 25 officers and 525 enlisted that served at Henderson Field on Guadalcanal. This was typical of the CASUs deployed to the South Pacific.

The Navy eventually planned to establish 70 CASUs. However, by the end of the war, only 69 CASUs had been placed into commission. Specifically, 10 CASUs were assigned to the East Coast, 15 to the West Coast, four to the Hawaiian Islands and 40 to Western Pacific locations.

The first handful of CASUs were assigned as follows. CASU-1 and -2 were at Pearl Harbor and Barber's Point in Oahu, Hawaii, respectively. CASU-3 was at Noumea, New Caledonia, while CASU-4 was placed at Puunene, Maui, Hawaii. CASU-5 was established at Naval Air Station (NAS) San Diego, California on 21 July 1942, and it remained at that location for the entire war. During 1942 and 1943, CASU-5 was the collection point for assembling, training and equipping CASUs as required for deployment to the Pacific. CASU-6 was at NAS Alameda, California, and CASU-7 was established at NAS Sand Point, Seattle, Washington. CASU-8 operated out of Kaneohe Bay, Hawaii, while CASU-9 was established in Suva, Fiji. CASU-10 was located on Espiritu Santo, New Hebrides, and, finally, CASU-11 was commissioned 22 January 1943 in San Diego, and, by mid-March it was operating out of Henderson Field, Guadalcanal. Fifty-eight more CASUs would be organized and deployed during the following two years.

"On commissioning day, 1 September 1942, ACORN (Red) One consisted of 44 officers and 1,347 men. It included the 14th Naval Construction Battalion (Seabees), a modified CASU (un-numbered), and a Base Headquarters section."[26] This joining of ACORN (RED) One with a modified CASU was the first operational deployment of an ACORN and CASU in World War II. Muster lists for this un-numbered CASU show that it was manned with 229 Sailors.

On 6 September 1942, ACORN (Red) One boarded the M/S Sommelsdyk in Oakland, California, and immediately proceeded to Port Hueneme, California, for additional cargo. On 9 September the Sommelsdyk was joined by the M/S Manoeran and they got underway for the western Pacific.

On 30 September 1942, both ships anchored in Noumea Harbor, New Caledonia, and offloaded cargo and a few personnel. On 1 December 1942 both ships arrived off of Guadalcanal and offloaded all remaining cargo and passengers. At this time there were no piers for unloading cargo on Guadalcanal. All cargo was loaded on small boats and delivered to the beach. Operating the boats for the unloading of ships was an initial responsibility tasked to ACORN (Red) One.

All ACORN (Red) One personnel immediately began working with the Marines to repair an emergency airfield adjacent to Henderson Field. On 16 December the first airplane landed on this repaired field. On 1 January 1943, ACORN (Red) One reported they were working on constructing permanent camp facilities, an airfield, roads, bridges and a hospital. They, also, reported servicing and repairing aircraft.

The requirement for aircraft services was rapidly increasing and became more than the 229 Sailors of the un-numbered CASU unit could handle. On 4 February 1943, "a chief in charge of forty-nine men was ordered to CACTUS for temporary duty in servicing and repairing aircraft."[27] CACTUS was the code name for Guadalcanal.

It is important to note that ACORN (RED) One arrived on Guadalcanal while the Marines were still fighting the Japanese on the island. On 9 February 1943, the Marines declared that the "organized resistance" on Guadalcanal had ended. Few were aware at this point "that from then on Japan would be on the defensive and the United States on the offensive."[28]

On 14 February 1943, the Naval Air Center Guadalcanal opened and began to relieve ACORN (RED) One of its airfield management duties. The Naval Air Center Guadalcanal was now responsible for all Navy housing, construction, food, and recreation. This was the normal transition, after airfield completion, permitting the ACORN and its Naval Construction

Battalion to move to the next location requiring airfield set-up and construction.

Naval Air Center Guadalcanal. From David Wilson.

On 15 May 1943, the 14th Naval Construction Battalion was detached from ACORN (RED) One and reassigned. The remainder of the ACORN personnel were distributed to other Air Center and airfield tasks. On 18 May 1943, the "temporary duty" augmentation manning of the Chief and the 48 men were returned to their original ACORN. Finally, on 9 June 1943 ACORN (Red) One was completely "dissolved in accordance with Commander Fleet Air Noumea Airmailgram 292202 of May, 1943."[29]

Guadalcanal was one of the earliest locations the Navy ever established for an ACORN - CASU organization. It was put together rapidly at a very remote location in harsh conditions. The arrival of personnel, supplies, and equipment was problematic. What was needed was often not available and regularly described as "it will be on the next ship." The Allied air force had been a piece-meal, patchwork of aircraft and aircraft repair men - they called it the "Cactus Air Force."

Part Three B - CASU-11 Action on Guadalcanal

On 18 March 1943, the men of CASU-11 knew they were finally offshore of Guadalcanal and they would begin their participation in the Solomons Campaign. This would culminate in the take-down of Rabaul. They had no idea how important they would be in the war effort in the South Pacific. The arrival of these aircraft maintenance and repair men would become vital in meeting the rapidly expanding missions tasked to Navy aviation on Guadalcanal.

Two days later, CASU-11 boarded small boats with their gear and were transported to narrow black sand beaches backed by hundreds of coconut palms and distant jungle-covered mountains. What would tomorrow bring? From Lt. Cmdr. Schlossbach all the way down to the most junior seaman, no one knew. However, it was certain that this story would unfold in a totally foreign environment of tropical heat and humidity, containing an unexpected set of dangers, noises and sounds unlike anything they had ever experienced before.

On 22 March 1943, CASU-11 assumed responsibility for the repairs of all Navy aircraft at Henderson Field, Guadalcanal.

Sat. 20th Nothing new. Sun. 21st Lashed up to go ashore. Mon. 22nd P.O. over ship watch. Tues. 23rd Same. Nothing new. Wed. 24th Came ashore. Rain like heck. Slept in tents. Had first bombing attack. Spent all night in foxholes. Licwindo sick. From the diary of Harry Hays, March 1943.

The SS President Polk was in Port 4, Guadalcanal, from 18-24 March 1943. What seemingly should have been an easy off-loading of CASU-11 was marked by interruptions from alerts of potential attacks and the rearranging of ships in the bay.

"We got in there and anchored in Iron Bottom Bay," O'Flynn said. "We didn't get the anchor set good until we had a submarine attack. We had to get underway and go back to sea again 'til the next day [when] we came back and off-loaded the ship."

According to the deck logs, Friday, 19 March, and Monday, 22 March, were the only days the men had to unload the ship without interruption. The men were under alerts on 20 March, 21 March, and 23 March. During a general alert or red condition alert, the ship's personnel reported to defense and fire stations. Battery (Guns) and communication stations were manned, and the remaining troops went below deck.

On 24 March 1943, the SS President Polk was underway, leaving its Sailors and cargo on Guadalcanal. Just before the ship departed, Hays got off at Lunga Beach.

Lunga Beach, Guadalcanal, was chaotic with multiple ships simultaneously sending small craft with mixed loads of personnel, crates and vehicles to the beach. When Little got to the beach, he found himself in a group of about 30 CASU-11 men, all with sea bags on their backs and no direction as to where to go. Down the beach, Little could see men struggling to drive their Jeeps. "The beach sand was so soft and fluffy; the Jeeps were having a great difficulty getting traction. They kept getting stuck," Little said. "Then the Beachmaster, in charge of the offload, noticed us Sailors standing around with our sea bags. Next thing I knew is he is ordering us to place our sea bags down on the sand to create a canvas road. Well, it worked, and we stood around watching Jeeps drive over our bags until we got told where to head for our campground. The Jeeps did no damage to my stuff."

Marines landing Jeeps on Lunga Beach, Guadalcanal. Marine Corps Photo.

Guadalcanal greeted the men with a tropical oceanic climate. The weather was warm and humid, and this was relieved by cool winds and abundant year-round rainfall. Temperatures rarely exceeded 90 degrees F, and rainfall generally averaged 120-140 inches a year, four times the amount Seattle receives each year. The rains were heaviest from December through April, with anywhere from two to six storms of hurricane intensity annually, usually in January or February, with wind velocity up to 100 miles per hour, and 5 inches of rain within 24 hours.

The result was a unique climate with air that was warm and humid, and either very still and moldy, or wet and windy. It was a dangerous climate for Sailors pulling and replacing aircraft engines, moving and loading ordnance, and doing the myriad other tasks required to keep aircraft continuously ready. Sailors were either knee deep in mud or exposed to the scorching sun and churned up dust with every footfall. Heat exhaustion was a perpetual threat to anyone working in these conditions.

Here are some of the sounds you might hear depending on where you were standing and what time of day it was: palm fronds clapping in the winds; chirping, cawing, and whistling birds; Jeep traffic slurping through axle-height mud; bulldozers starting up; aircraft engines starting and stopping; swearing; aircraft taking off; distant bomb concussions; and gunfire from both small arms and big anti-aircraft guns.

After dark, it is almost guaranteed that the men would have heard the disturbing sounds of Washing Machine Charlie. This twin-engine Japanese aircraft made noises produced by out-of-sync engines that would wake the dead. Its load of small bombs were mixed with whistling bottles, created for the purpose of making disruptive noise and driving the men into foxholes. Marine Robert Leckie described him in his book, *Helmet for my Pillow,* as, "the nocturnal marauder who prowled our skies. Charlie did not kill many people, but like MacBeth, he murdered sleep."[30] "The Charlie character became a fixture over every Pacific battlefield but he was best known for his flights at Guadalcanal."[31]

And if Charlie could not make the flight then "the single engine Louie the Louse, a Zero floatplane, would assassinate sleep with sporadic after-dark bomb drops onto"[32] Guadalcanal. He generally performed his night-time ballet by making repeated passes over the Guadalcanal airfields, more for noise than bomb drops. He often dropped flares, and this was often followed by bombardment from a nearby Japanese battleship. Washing Machine Charlie and Louie the Louse only flew at night, ensuring no one on the ground could see them well enough to shoot them down or launch a plane to attack them. This nightly disturbance was eventually brought to an end, first by the installation of radar on Guadalcanal that could help guide search lights and ground-based anti-aircraft guns onto these night-time targets and second, by the Marine Corps deployment of Marine Night Fighter Squadron VMF(N)-531 operating night time interceptor aircraft from an airfield on the Russell Islands.

There was never a moments' peace. Evening sounds also included the ever-present buzz of flies and mosquitos, the clicking of the feet of tiny mice as they scampered across tent floors or land crab pincers as they explored everywhere, and maybe, the loud snoring from the adjacent tent. Occasionally, Japanese submarines or destroyers cruised by Guadalcanal, lobbing rounds hoping to destroy a parked airplane or hit a bomb storage area and cause secondary explosions.

The following camp map was found in files found at the National Archives in San Francisco, Record Group 313, Records of the Naval Operating Forces, Naval Air Base Guadalcanal Island, Solomon Islands 1943-1946.

This is Lunga or Lungga Point. It is the portion of Guadalcanal where the Marines commenced their efforts to take a Japanese airfield. This became Henderson Field and established an Allied presence in the South Pacific. Kukum Field, Henderson Field and Lunga Field are shown from left to right. Henderson Field was the center of CASU-11's early war efforts. Numbered camp areas were established around the three airfields.

CASU-11 lived at Camp 53. The camp, located at the tip of the arrow, was not far from the beach on the eastern side of the right most stream of the Lunga River. The aircraft parking area for Henderson Field was about a mile south of Camp 53 along Road Number 48. Both sides of the road had thousands of coconut palms planted in equally spaced and uniform rows.

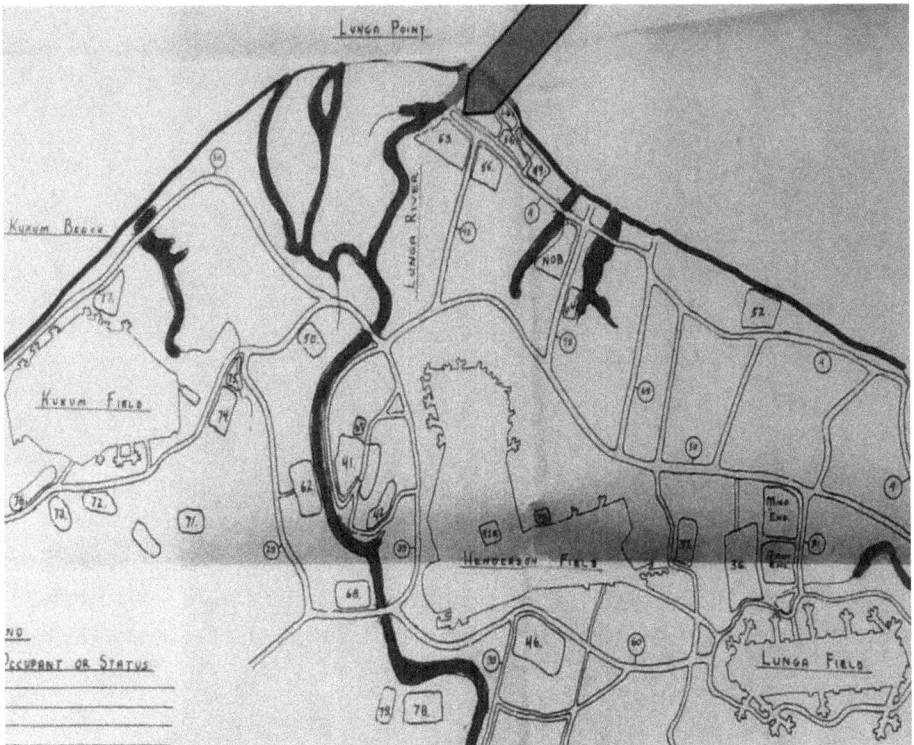

Camp Map found in National Archives Regional Office, San Francisco.
Photo by author.

O'Flynn remembered Guadalcanal being, "a very mountainous place, and they had two or three rivers there. We used to bathe in one of them. Talking to the natives over there, they had some Chinese with them in the production of coconuts. 'Cause all the island's beach areas were all coconut trees. Talking with some of them, the natives would go to this one river there, it was about a quarter of a mile from our tent area. They would go there every day to take a bath, men, women, children, all. I was asking, you go there to take a bath? They said yes, everybody strips off and goes and washes themselves, they says. Somebody asks, he says, do they pay any attention to each other, the sexes? He says it's not a sin to see a naked body, it is a sin, if you goad on it."

Guadalcanal Bathing 1943. Marine Corps Photo.

The men of CASU-11 learned quickly that the Lunga was home to crocodiles, and as a result the bathing process consisted of 10-15 naked men with towels and soap and shoes, followed by one or two dressed men with rifles. Crocodile watch was mandatory for the Lunga!

Aviation Housing Report Guadalcanal 1 Sep 1943. Photo by author.

The above housing reports dated throughout 1943 and 1944 show CASU-11 remained in this location, and O'Flynn verified that he was housed in Camp 53. The line for Camp number 53 reads as follows: As of 1 September, 1943 CASU-11 occupied this site with 22 officers and 557 enlisted. Housing was divided into officer and enlisted sites. The officers had eleven floored tents, and the enlisted had 14 unfloored and 162 floored tents. The total capacity of the tents was 704 occupants. Since CASU-11 was not that large, some of the men had less than four men in their tents. This 1 September 1943 Aviation Housing Report was also found at the National Archives in San Francisco, Record Group 313, Records of the Naval Operating Forces, Naval Air Base Guadalcanal Island, Solomon Islands 1943-1946.

CASU-11 tent city after rainy day. From John McAteer.

According to O'Flynn there were four men to a 10 foot by 10 foot tent. His was floored. "Didn't have any hanging clothes lockers, they had foot lockers. Of course, before the tour was up, all the clothes we had before, the uniforms, they were all shot down. [All clothing brought over on the SS President Polk quickly rotted in the humidity, mold and working conditions on Guadalcanal.]" O'Flynn's final words on the living conditions: "And, oh yeah, Navy had to pay for their clothing, shoes and everything. Shoes cost $6. Dungaree shirts were about $1.75 and pants were about ... a little bit more, I think."

Marine pilot Foster, in his book *Hell in the Heavens,* said upon his arrival at Henderson Field, "We were taken to the Hotel De Gink - nothing more than a Nissen hut - a concoction of metal, wood and mosquito netting shaped like a hot dog that has been split lengthwise, with only the upper half being used. Our back yard consisted of a plantation of coconut trees replete with an occasional nut. But mostly the tops of the trees were frayed from the battle that had raged a year before."[33] The Hotel De Gink was located on the edge of Camp Site 53. By September 1943 when Foster arrived, the camp had added Quonset huts in among the floored tents. Hotel De Gink was for officers only.

CASU-11 Tent with residents, L – R Broughton, Flora, West. From Durward West.

Astor shared three related stories in his book *Semper Fi in the Sky*, "And there was no electricity (early on) so we had no music or lights, just lanterns. I had a tooth problem and the doc, not a dentist, had to drill using a drill powered by me pedaling like a bicycle as he drilled. The faster I pedaled, the faster the drill went. It wasn't very pleasant but I think I may still have that filling to this day." And, "we had movies on the 'canal. I remember we sat in the rain and watched them, even if we had seen them before."[34]

Packing boxes were used for tent decks, and it was not until August 1944 that enough material was available to deck all tents on the island. Electric lights in tents were not installed until September 1944. These improvements occurred after CASU-11 had departed Guadalcanal. Flashlights and kerosene lanterns were issued to the men of CASU-11. Even mail delivery was a challenge. The lack of specific storage space, transportation to deliver and personnel to sort resulted in partial or total loss of many bags of mail as late as March 1944.

Wukovits' book, *Black Sheep,* shares this additional tent camp description, "mosquito netting helped reduce the influx of mosquitos and other insects, but nothing completely eliminated the nighttime hordes that infested the camps. Near unbearable heat and humidity smothered them around the clock and dampened their clothing. Hour-long daily torrential rains further heightened the misery, and each time a pilot took to the air they had to wonder whether they might have to land a damaged aircraft during one of the heavy downpours."[35]

Thurs. 25th Work, work, work. Another alert. Fox Hole again. Fri 26th Same. Fox Holes again. Everyone squared away but us. Next to the 9th Div. Sat. 27th Bombed again last night. About 10 Negroes killed. Fox holes for over 3 hours. Digging fox holes of our own. Have choice of CM 1/c or AM 1/c. Which? From the diary of Harry Hays, March 1943.

CASU-11 Men taking a photo break from digging foxholes. From Durward West.

Henderson Field aircraft parking and service area to left of main runway. Lunga Field is to the right. Navy Photo.

CASU-11 spent most of its working hours with aircraft parked in four loosely designated areas located just off the west end of the main Henderson Field runway and on both sides of Lunga Field. In the center of the above picture it is the broad area with "chicken feet" between CASU-11's camp on the north coast and the main Henderson Field runway. Also notice the "chicken feet" parking areas around Lunga Field.

Squadron airplanes were generally parked in the two zones closest to the runway, and the men of CASU-11 worked primarily in the outer three zones overlapping with the squadrons in the minor maintenance zone. These zones were designed to ensure that the aircraft were spread far enough apart to prevent a single Japanese bomb from damaging several airplanes at once. They were also close enough to the runway to permit a quick launch. The first zone of the four zones consisted of squadron ready airplanes. The next zone was aircraft undergoing regular minor maintenance; such as 30, 60 or 90 flight hour checks. The third zone was airplanes needing significant repairs and/or major overhauls; such as an engine replacement. Aircraft that were beyond repair and had valuable spare parts were left in zone four on the edge of the jungle.

Runways were always oriented into the prevailing winds. These winds gave ascending aircraft added lift and those descending some natural braking. At Henderson Field these winds were north-easterlies. Aircraft work areas also accounted for the prevailing winds. When the men were testing the engines, they needed to ensure that the winds did not blow dust and dirt spun up by spinning propellers into an open cockpit. The placement of the aircraft away from the blowing dust was essential when the men were cleaning and reloading armaments or maintaining sensitive radio equipment.

Little remembered "Henderson Field was mud and dust. Alternating rain and sunshine, the rain creating mud that was knee deep, and then the sun was drying the top the ground off, and the dust was blinding. You couldn't see nothing for dust."

Bergerud shared what these conditions meant, "Dust was a maintenance nightmare. Aircraft of the era were extremely complex and extremely vulnerable to erosion caused by dust simply because the planes' cooling apparatus sucked in air at all times and circulated the mud and dust into the inner workings. Likewise, dust was the bane of instruments."[36]

Consequently, CASU-11, with almost 600 highly trained mechanics and aircraft service personnel, provided at least ten men for every aircraft they repaired or maintained. This level of manpower was necessary to keep complex World War II airplanes functional in the demanding environment of Guadalcanal.

During a workday many aircraft moved from one workstation to the next. Engine replacements and testing, gun aiming and test firing, and wing removal and replacements required specialized equipment. These included large tripods with chain falls to lift heavy engines, jacks to lift an airplane into its normal level of flight geometry for proper gun aiming and a metalsmith shop equipped to bend sheets of metal into replacement fuselage parts.

Prior to the arrival of CASU-11 airplanes were refueled with hand pumps that were used to transfer fuel from 55-gallon steel drums into aircraft fuel tanks. The men had to manually roll the steel drums to the aircraft. By the time of CASU-11's arrival there were numerous fuel trucks available with diesel powered pumps. The trucks drove to the airplanes for refueling.

Finally, for everyone's safety special ordnance handling areas were established to store, assemble, and load bombs onto the aircraft. This generally occurred in squadron designated areas close to the runway and was completed shortly before takeoff.

Initially all work on the airplanes was done during daylight. A few months after CASU-11 arrived diesel generators became available and the men could work at night.

While the men of CASU-11 were busy working, Henderson Field was busy with many takeoffs and landings. Weather reconnaissance flights flew a couple of times a day and flights arrived daily from New Caledonia and Espiritu Santo with new personnel and supplies. Long range surveillance flights flew in search of enemy locations and enemy strike response fighters were always poised for launch. These were intermixed with maintenance test and new pilot orientation flights. If a typical Navy air group started with 100 airplanes, within a few weeks of action over wartime Guadalcanal 45 of the aircraft would be in maintenance and repair. This would leave the air group with only 55 mission ready aircraft, a number that would continue to slowly decline due to accidents and losses from enemy action. The only relief from this was the arrival of new aircraft.

On the web site "You Tube" there is an interesting and informative movie on the 17th Photo Reconnaissance Squadron. Their camp was at site 49, just a few hundred yards east of CASU-11. As you watch the movie take in the surroundings and the quick change from mud to dust. CASU-11 experienced these same conditions. The movie is at: https://www.youtube.com/watch?v=88p9z6J-CiI&t=533s or search for 17th Photo Reconnaissance Squadron.

An "alert" signaled enemy aircraft in the area and a "raid" signaled that bombs were exploding on the ground. The formal warning system included two warnings: Condition Yellow (enemy planes will be overhead in 30 to 45 minutes) and Condition Red (enemy planes are overhead, ready to bomb; proceed to foxholes immediately.) However, there was also Condition Black, which meant an invasion across the beach was imminent. Condition Very Red meant bombs are dropping now, now, now. Finally, Condition Green was all clear.

Henderson Field was not the only site that sounded alarms for "Condition Red - enemy aircraft overhead." The alarm at Henderson Field was for all personnel within hearing distance, which included CASU-11 whose camp site was barely a mile away from the field. However, O'Flynn remembered that actually, "The sirens we heard was from an Army unit maybe a quarter of a mile away, the siren you could hear at a quarter mile very well. They would then get on the old speaker and say … 'we are having a red alert,' and when it was over with they would say 'air raid finished for the day or for this time.'"

Computing the actual number of bombing alerts and raids has been difficult. Wilson noted important events on a card he kept in his seabag. His notecard is provided below with several raids during the period March through July 1943.

Notecard. From David Wilson.

According to AM1c Wilson's notecard: Major air raids occurred on 24 and 27 March. 7 April had a daylight raid followed by raids on the 17th and 19th. May started with a raid on the 13th that lasted both day and night. There were raids on the 19th, all night on the 20th and on the 24th June. The 16th had a noteworthy daylight raid, which turned out to be the largest raid of the war on Guadalcanal, followed by a raid on the 23rd. Raids occurred on 11, 16, 17, 19 and 20 July 1943. It's a simple notecard that powerfully reveals the events that happened those first few weeks on Guadalcanal.

The last page of Hays' journal lists the numbers of alerts and raids by month. March and April 1943 listed 20 alerts and 6 raids. Using other sources, during the first six weeks that CASU-11 was on Guadalcanal, Henderson Field set condition red 10 times during the last half of March 1943 and 26 times during April 1943. Whether the real number was 26 or 36 for this six-week period, it was still a frightening number. Here are O'Flynn's words about the bombing: "That was a nightly thing for a while, for months, then it got to where it was only a couple times a week, or 3 times maybe a week the last few months we were out there."

Frequent alerts meant the Sailors of CASU-11 made frequent trips to their foxholes. "Oh yeah, every tent had a foxhole," O'Flynn said. "One time along there, before they quit bombing us all the time, [we] had a bunch of Marines move in next door to us. So several of us went over there, talking to the Marines. Guy said, 'What you guys do for fun?'" O'Flynn recalled his group replied, "We dig foxholes!" According to O'Flynn, the Marines replied with, "These yellow-bellied characters won't hurt us Marines." The Marines probably should not have tempted fate, and O'Flynn continued: "So that night they [Japanese] showed up. Next morning I went over there, and, well this place looks cluttered up, I said. They [the Japanese] had hit their supplies, scattered them all around, and they had two vehicles, knocked them badly out of commission. Two of the guys got shrapnel hits on them; they took them away. So that tickled the fool out of me. So you dag-blamed Jarheads, they tell you in boot camp, they can't kill a Marine, but they do. As I was walking back to my camp that was funny, but it was really life-threatening. It wasn't a laughing matter at all. But my mind was sort of jammed up. Anyhow, all the Marines, that next morning, were too busy digging foxholes to talk."

Every airport is unique in one way or another. Guadalcanal was no exception, the singularity of the threat axis, that is, the direction the enemy airplanes came from, was always from the north northwest. The Japanese had several airfields in the northern Solomon Islands that sent small raids, six to ten aircraft, all hours of the day and night. Bigger raids, with greater than 20 aircraft, came out of Rabaul and Kavieng. However, "the distance from Rabaul and Kavieng forced the Japanese to leave early in the day, usually arriving over Henderson Field between eleven and twelve o'clock in the morning. Moreover, the flights invariably arrived from the northwest, often after forewarning from Coastwatcher stations and radar. The long distances, American awareness of impending attacks, and inadequately armored Japanese planes gave the defenders an important advantage, which they used repeatedly to ambush the Japanese."[37] Altitude was everything, so early warning from Australian observers (Coastwatchers) spread up and down the length of the Solomon Islands was critical. Coastwatchers provided advanced notice that Japanese arrival could be between 15 minutes or an hour and a half. Typical warnings were about an hour, and even with this amount of time, Allied aircraft would race to get to altitudes above the Japanese. The resulting aerial tactics were designed to literally slash downward through the Japanese bombers, trying to shoot one or two bombers out of the sky. Then quickly get down to Henderson Field because the Japanese did not have enough fuel to hang around and fight. "After slashing through the flights of bombers, the Wildcats dove to safety at Henderson Field. Moreover, American airmen shot down could be rescued. Downed Japanese pilots were normally lost at sea or captured."[38]

Additionally, "Poor ground to air radios limited the intercept zone to an area ten to fifteen miles from Henderson Field. In the sky above the field within that distance, the Wildcats' dive and run tactic was very effective and compensated for the limited [size of the] interception zone. The resilient airplane [The Wildcat] in the hands of capable pilots shot down or damaged a disproportionate number of enemy aircraft."[39]

Sun. 28ᵗʰ Another alert. Fox holes for 1 hour. Trying to help Metalsmith and work as carpenter. Flight skins is reason. Trucks taking gas barrels away, thanks heaven. Have 8 hours of watch. Have to work this P.M. From the diary of Harry Hays, March 1943.

The term flight skins has two meanings. It may refer to extra monthly pay for anyone who has flying responsibilities. It may also refer to the skin of an airplane. Hays had previously written he had a choice of becoming a "CM 1/c or AM 1/c" which means Carpenter's Mate First Class or Aviation Metalsmith First Class, and now he says "Flight skins is reason." This is confusing, however, an AM 1/c could be paid for a little flying to check out their work on the metallic skins (fuselage) of a damaged airplane. In the end Hays chose to become an Aviation Metalsmith.

Mon. 29th Work. No alert. Tues. 30th Same. Wed. 31st Same. Still haven't had time to dig foxhole. From the diary of Harry Hays, March 1943.

While Hays was deciding whether to become a carpenter or a metalsmith, O'Flynn was working at the motor pool. [We] "worked on engines, maintained fleet of support for the Air Station, Strip. We had two little old boats, but we never used them. We had two DUKS, they were a jeep you could run in the water. Our main project was to keep the fleet running the trucks and the jeeps, carryalls and all those things we used. We had a couple of bulldozers, a couple of cranes, and that stuff we had to keep in top-notch shape."

When CASU-11 arrived on Guadalcanal, Henderson Field was already in operation. Between the two fighter airfields and the bomber strip, there were almost 300 aircraft. Morison in his book *Breaking the Bismarcks Barrier,* reported that after a day of fighting off Japanese aerial attackers over Guadalcanal "three hundred planes of the armed forces roosted nightly on the Lunga plain; nearby encamped hundreds of men to serve them with bombs, bullets, fuel and overhauls."[40] The largest part of those "encamped" men were assigned to CASU-11 and they knew considerable aircraft maintenance work awaited them.

CASU-11 was not alone, three other CASUs were also operating in the South West Pacific at this time, CASU-3 at Tantouta, Noumea, CASU-9 at Nandi, Fiji, and CASU-12 Efate, Vanuatu. This presence would grow rapidly. Six months later, three more CASUs had arrived and were operational, CASU-8 was in Segi, New Georgia, CASU-10 in Espiritu Santo, Vanuatu and CASU-14 at Munda Field, New Georgia. As Allied forces continued to take islands and build airfields, more aircraft and personnel arrived to man the airfields, each success pushed the Japanese further north. By March 1944, a year after CASU-11 arrived on Guadalcanal, there was an additional six CASUs deployed to the area. CASU-19 on Russell Island, CASUs 13, 39, 40 and 41 at Espiritu Santo and CASU 15 on Efate. At this point over 750 aircraft were spread across these locations with several aircraft carriers battle groups contributing another four to five hundred aircraft as required to continue working up the Solomons.

During the 17 months CASU-11 was on Guadalcanal, they had a monthly average of about 30 miscellaneous aircraft of various types under their management for repairs and maintenance. The Commanding Officer of CASU-11 was responsible for these aircraft. This count did not include the quantities of aircraft that dropped in for irregular repair and maintenance, a number driven by the variability of war.

CASU-11's primary responsibilities were to properly maintain all the aircraft under their charge and return all damaged aircraft back to flying and fighting condition. They kept aircraft in the sky and even stripped severely damaged aircraft for parts to rebuild other aircraft. And they did it with great speed. An additional onerous task that never did end for CASU-11 was the receipt and preparation of newly arriving aircraft, from the factory. A new aircraft arrived in multiple parts and numerous crates. Sometimes the wings were separate and required mounting onto the fuselage, other times the wings just unfolded. The propeller had to be attached to its shaft, the engine drained of preservatives and refilled with lubricants. Then, and this was the worst part of this job, came the removal of all anticorrosive coatings from all external surfaces of the aircraft and from the guns. Cosmoline, a heavy petroleum or grease, or para ketone, a lighter petroleum product, was used by the military to protect its equipment from rust and corrosion during shipment. With time and heat, it could be very difficult to remove. Freshly applied cosmoline had a grease-like viscosity and wiped nearly clean with a rag, leaving only a thin film behind. Older cosmoline which has had lengthy air exposure could solidify leaving behind only the waxy remainder, something weeks in the tropics only made harder to remove. Consequently, it took 200 man-hours to get an entire plane out of the box, reassembly it, remove the anticorrosive coating off all surfaces and then tuned it up for flying. This was CASU-11's role, nonstop, practically every day.

Appendix F describes the variety of aircraft that CASU-11 repaired and maintained during the war.

April 1943

CINCPAC *Operations in Pacific Ocean Areas* report for April 1943: During April our activity in the South Pacific was largely devoted to attacks on Vila-Munda, with isolated strikes farther north, and to continued development of bases in Guadalcanal-Tulagi and the Russells. The Japanese, continued to strengthen their defenses and increase their airfields of which there were some 14 in the Solomons-Bismarck area by the month's end. In April our aircraft dropped about 400 tons of bombs on enemy airfields from Vila to Rabaul without noticeably affecting enemy air activity except from Vila-Munda fields.

At 1400, 7 April, radar on Russell Island indicated a bogy 300 degrees True, 148 miles from Henderson Field. All available fighters took off to meet the enemy, 76 getting into the air and 56 making contact. By 1435 about 50 enemy planes were orbiting between the Russells and Savo, with another group approaching. At 1459, 8 of our fighters over Savo were attacked by large numbers of enemy Zeros. For the next half-hour there were air duels from Savo east to Florida and down the coast of Guadalcanal as far as Rua Sura Island east of Lengo Channel. Our fighters were engaged so closely by enemy Zeros that few had opportunity to attack dive bombers until they had dived on our ships. (Other sources reported this Japanese strike had 110 Zeros and 67 Val bombers.) Operations for the remainder of the month were more or less routine bombing attacks on Vila-Munda.

Thurs. 1ˢᵗ Same. Had to pay native $1.00 to wash clothes. Got stuck. Lousy work. Got $5.00 Beer card. From the diary of Harry Hays, April 1943.

When CASU-11 arrived on Guadalcanal, the native population was estimated at 15,000, while the military population was 50,000. Many natives were offered a variety of jobs including unloading ships and repairing roads and runways.

Fri. 2nd Can't change to carpenter. Still work as such, though. Sat. 3rd Same. Sun. 4th Changed tent again. P.O. of 8 to 12. Made application for officer's training. Mon. 5th Pay day. Have $112 coming. Nothing new. Had 8 to 12 last night. Poured rain. What a mess. From the diary of Harry Hays, April 1943.

Hays indicated that he was the "P.O. of 8 to 12." This means he was the Petty Officer in charge of the CASU-11 men assigned to the 8:00 AM to 12:00 noon watch. This was probably a security watch responsible for monitoring the CASU-11 camp site.

Over any tropical island, thunderstorms are typical almost every afternoon. Cumulus clouds build up during the heat of the day and dump rain, sometimes intermittently and other times in a massive downpour. Typically, the clouds would rain out and be gone shortly after sunset. It was so hot between rain showers that puddles would evaporate rapidly. It could be dry and dusty one minute, and muddy the next. This alternating mud and dust was tough on operations. The mud impeded the movement of heavy vehicles, and the dust blew into eyes and precision instruments. Taylor, in his book, *The Magnificent Mitscher* noted, "In April 1943 especially, there was rain, rain, rain. Sometimes three inches would fall in a morning, and they would have to wade to meals, sloshing and slopping in the mud. The Lunga [River] was in full flood and huge tree trunks tumbled slowly in the froth."[41]

Little remembered "Guadalcanal was mud and coral, and so much mud that when it rained you was wading in it up to your knees. And we would go and scoop up the coral off the beaches and was hauling it, not for us but for the Army, hauling it and they built us roads out of it. I don't remember it being so slippery, but maybe it was, and then another thing was we had so much mud in it, it was slippery anyhow."

One thing that wasn't found on a coral atoll was gravel, so the troops used what was in abundance: coral. Bennett further explained in her book, *Natives and Exotics,* that, "most used was live, blasted away from its foundations in the reefs and lagoons and dredged up to be crushed and compacted into a surface for runways. If stabilized by being moistened with water or sealed, coral made a hard surface. On Guadalcanal's extensive roads and runways, the demand was ongoing, so the main (coral) borrow pit at Kukum, almost four miles west, across the Lunga River, from Henderson Field, had to be reserved solely for airfield maintenance."[42] The Kukum coral quarry was so large that, at war's end, new and used equipment was tossed into its cavernous hole and then covered over, just to get rid of it.

Crushed Coral Road. Navy Photo.

Merillat wrote, "The only good thing about the rain is that it keeps Tojo out of the skies. No, there are two good things: it also keeps the rats down."[43]

Little remembered, "When I went up to Guadalcanal there wasn't any flies to amount to anything, very, very, few flies but mice, oh my there was mice." The mice, actually Guadalcanal rats, Uromys porculus, were a nuisance and a problem to be solved by the men. "What we had was this board run up to the top of an open barrel. Then we had a real round stick with a cookie tied to the middle, extending across the top of the barrel. The mouse would try to get out and get that cookie, and that round stick would roll over and drop him down in the barrel of water, where they would swim all night. It was noisy, but when we mixed in gasoline, that put the end to them and ah, each morning you had to clean it out because if you let very many of them in there, they would start stinking with that high heat that they had there in the island."

Tues. 6th Another raid. Caught us with our pants down. Right after movie. I was 50 yds. from fox hole. Fell flat and prayed. 3 bombs hit not over 200 yds. away. Ran for fox hole. We got one plane. Shrapnel piece or some part of plane fell through tent 2nd from ours. Too close for comfort. Hope we all come out of this alright. Wed. 7th A daylight raid. Supposed to be 150 to 200 Jap bombers on way here. We intercepted them. Got 11, lost 1 but saved pilot. We watched dog fight. Saw 2 planes fall. One in flames. One in smoke. This is the real thing. From the diary of Harry Hays, April 1943.

Dog fights were something every member of CASU-11 witnessed as the war played out directly overhead. O 'Flynn told the following story, "This one [aerial fight] had a F4U plane, an American plane, it was one of our main fighters in those days. Behind him was a Jap Zero firing on the F4U, and behind it was an Australian firing on the Jap. One of the guys that was near me, told me come over here, you will get a good look at this. I don't want to see that! It was just like in the movies. So I went over, and he said, you see, they are on about a 15 degree down angle; they had better change their course or they are going to run in, run out of air space. So you could see for miles, and this was probably only a mile out. Here comes the F4U, splash, hits the water, flew apart. Right behind him, a Jap flew, hit the water, and blew apart. Behind him, the Australian hit the water and tore apart. We got to talking about that: so they were so involved in what they were doing that they didn't think about killing their selves, just had other things on their minds."

Thurs. 8th Two alerts last night, but no trouble, except loss of sleep. Moral[e] busters, these raids are called. Fri. 9th Work. No alerts. Received 12 letters. Application has to have 4 copies. Nuts! Sat.10th Army show tonight. Going to write letters instead. Talk of invasion going around. Hope it is scuttlebutt. Island has been under condition Black for a week without us knowing. That's invasion condition. Condition red is air raid, condition green is all clear. If there is an invasion, we won't have a chance. All green hands and right at the mouth of the Lunga River where the Marine's bloodiest battle was fought. Also right on beach where they would have to land. Dear God take care of all of us. From the diary of Harry Hays, April 1943.

While CASU-11's primary role was to maintain aircraft, they were also a critical part of the defensive shield protecting Henderson Field and other U.S. military assets in the Solomon Islands. The Japanese attacked Guadalcanal 36 times during the first 44 days that CASU-11 was on the island, from mid-March through the end of April. Consequently, every day CASU-11 had to have every available Allied airplane ready and flying over Henderson Field to protect the military on Guadalcanal. Over 300 aircraft could be put into the air late in March 1943, and by June the total would exceed 450.

Among the numerous types of aircraft maintained by CASU-11, there was one remembered by all hands - the Scout Bomber Douglas (SBD) Dauntless Dive-Bomber. It would prove to be the Navy's most successful dive-bomber of the Pacific War. "The Dauntless weighed more than four and a half tons when it was fully loaded for combat. It was capable of carrying a lethal thousand-pound bomb for attacking capital [large] warships, or smaller loads for attacking land-based targets. Each time one of these dive-bombers returned to base someone would always remark how rugged this plane was."[44] CASU-11 metalsmiths repeatedly were amazed by how many bullet holes these planes accumulated and still made it home. With an unremitting threat of Japanese invasion, the men of CASU-11 took excellent care of the SBDs because they, especially, could keep the enemy ships away from Guadalcanal's beaches.

Additionally, throughout the Solomons campaign, Marine and Navy bombers based at Henderson Field played a critical role in both the naval and land target equation. In the waters surrounding the Solomon Islands, it was rare to see Japanese warships larger than a destroyer. Therefore, aircraft were tailored to demolish air bases and destroy smaller vessels. The two bomber aircraft employed by the Marines and the Navy were the Douglas SBD dive-bomber and the Grumman TBF torpedo-bomber. SBDs and TBFs played a significant role in the Solomons victory. They demonstrated their value by being small, durable, fairly nimble and accurate when attacking a discrete target, such as a ship or airfield. And they were able to operate from almost any small, jungle airfield.

SBD near Henderson Field Tower. Navy Photo.

"The best-remembered description of the Dauntless was an acronymic play on its designation. "Slow But Deadly" was the way many SBD pilots thought of their airplane, and certainly it was accurate, for the Dauntless was the most successful and the best dive bomber of the war. Its inherent stability made it the steadiest possible sighting platform in a dive, and its light control responses made corrections easy when lining up a target."[45]

"The SBD became a magnificent weapon during a magnificent moment in history, as no aircraft did more to bring Allied victory in the Pacific war. But its moment of glory was brief... As the war progressed Japanese antiaircraft became so deadly that even a dive-bomber formation was in peril."[46]

Sun. 11th Had drill for invasion. Went to Mass. Work. Barker cried a bit last night. Homesick. Had alert but no bombs. Feel lousy. From the diary of Harry Hays, April 1943.

Speculation ran rampant about what might happen next. An invasion was the most frightening possibility. At this critical point in the war, this was understandable as the Japanese were throwing everything they had against Guadalcanal. The Japanese knew if they were ever to establish permanent control over the western Pacific, they could not permit the Allies to remain on Guadalcanal.

Mon. 12th Went to sick bay. 104 fever. Probably malaria. There's lots of it around. Tues. 13th Still feel punk. Had 5 alerts last night but no action. Hirohito's son was killed on the 13th. We're expecting it today. My first malaria test was negative. Have 102 fever today. From the diary of Harry Hays, April 1943.

While on Guadalcanal, Little and O'Flynn were infected with malaria three times. On the third occurrence O'Flynn was evacuated to the Naval Hospital New Caledonia for almost a month. O'Flynn's daughter, Jann, shared with me that, "my father was saved by one of his tent mates because he recognized that Dad had malaria really bad and literally dragged him to the medical tent." Shortly after getting to Port Hueneme, California, and about three months after leaving Guadalcanal, O'Flynn had another bout of malaria. The malaria mosquito on Guadalcanal was five times more problematic than a Japanese Soldier hiding with his gun behind a coconut tree. For every Allied service member the Japanese killed or wounded, the mosquito sent five men to the nearest infirmary for malaria treatment.

When the Marines arrived on Guadalcanal to wrest it back from the Japanese, they paid little attention to mosquitoes and the associated malaria. In fact the "attitude was expressed by one high ranking officer who said - we are here to kill Japs and to hell with mosquitos."[47] It was not until leadership realized the five to one ratio of malaria over combat casualties that they changed their thinking regarding this disease. Usual malaria symptoms included fever and flu-like illness, including shaking chills, headache, muscle aches and tiredness. Leckie described symptoms for a second illness called "bone-cracking" malaria, the malignant kind that bakes your body in an oven and stretches your bones on a rack."[48] What Leckie had described could also be "Dengue Fever." It was very similar to malaria, comes from a mosquito bite and definitely presents the feeling of your bones cracking.

The solution to the malaria problem was twofold, force the men to take their anti-malaria medicine and combat the mosquitoes. The anti-malaria medicine was Atabrine, and the doctors decided the best way to distribute it was in the chow lines. However, it was quickly noted that the men were not inclined to take their pills - the ground was littered with yellow tablets. Making the rounds of the men's gossip network were stories of Atabrine being a poison, causing sex problems, and/or staining the skin permanently yellow. Leadership finally had to place the pills into their men's mouths and watch that the pill was indeed swallowed.

Combating the mosquitoes took manpower. When CASU-11 arrived at their assigned campsite, they could see the ongoing Seabee campaign to reduce the mosquito population. Mosquito larvae cannot survive in moving water, so every bit of standing water was either removed, filled in, sprayed with oil, or trenched to flow to the sea. Discarded oil drums were welded together to create long pipes that could be used to drain the swamps through the sand dunes to the sea. One report stated that more than 50 miles of ditches were dug by hand labor, dragline or blasting during this effort to eliminate the mosquitoes. By August 1943, conditions on Guadalcanal were greatly improved. However, management still felt the need to publish the following letter to be distributed to all Guadalcanal personnel:

Dated: 27 August 1943
From: The Commander South Pacific.
To: The Commanding General Guadalcanal.
The Commander Advanced Naval Base Guadalcanal.
The Commander Air Center Guadalcanal.

Subject: Malaria Incidence at Guadalcanal

1. The elimination of malaria amongst Allied Forces on Guadalcanal is a matter of utmost necessity if this island is to be of value as a military base. Reports show a continued high rate of incidence of this disease. As more men become infected, they act as a seed bed of this disease which may increase by geometrical progression.

2. Mosquito Control measures have been instituted and are being vigorously prosecuted. However, reports and observation indicate a continued disregard by personnel of elementary measures of individual protection which are in themselves the most important factors in combating malaria. The following excerpts from recent reports of Malaria Control Units are quoted:

"Malaria discipline: Continued violations are being noticed in the wearing of proper clothes by troops late in the afternoon and after dark. Frequent visits have been made to check on picture show attendance in proper uniform but organizations continue to allow men to go about after 6 P.M. without shirts or in shorts."

"Malaria discipline: is still lacking on the part of many units, especially in the wearing of clothes at night."

3. Responsibility for malaria discipline and vigorous prosecution of anti-mosquito measures within camp sites rests with unit Commanders. Recommendations and instructions in these measures may be obtained from Malaria Control Units.

4. A continued high rate of Malaria in any organization shall be made the subject of an investigation by the Commanding General of the island or by the Commander Advanced Naval Base Guadalcanal who will take appropriate action to prevent the continued wastage of manpower.

Signed   J. F. SHAFROTH
Deputy Commander
South Pacific

After the war was over, it was calculated that 60,000 of the 100,000 men who caught malaria in the Pacific Theater were on Guadalcanal.

This report of Hirohito's son being killed was incorrect. Emperor Hirohito had five daughters and two sons. All seven survived the war and son, Akihito, became the next Japanese emperor after his father's death in 1989. Other CASU-11 members heard the thirteenth was always a bad day because General Tojo's son was killed on the 13th. This, too, was incorrect. However, rumors have a life of their own, especially in war time.

Wed 14th One alert no action. Relieved the pressure a little. My fever 99.8 2nd test negative. Thurs. 15th No alert. No bombing 3rd test negative. Feel good. Fri.16th Nothing new. Sat. 17th Had another raid last night. About 45 minutes in fox hole. Boy, it's a funny feeling when you hear those bombs coming down. You have to just pray they don't land too close. Sun. 18th Palm Sunday. Went to Communion. Bishop said Mass. Been on Island 35 years. Very old. Mon. 19th Longest raid yet. We got at least 3. Saw one come down in flames. Broke in half. About 5 hours with helmet on. Wish I was home. From the diary of Harry Hays, April 1943.

Bomb Hit near Henderson Field. Navy Photo.

Two famous aviators spent time on Guadalcanal during the period CASU-11 was present at Henderson Field. One was a military pilot, and the other was a civilian pilot. They were Major Gregory "Pappy" Boyington, U.S. Marine Corps, and Mr. Charles A. Lindbergh, consultant with United Aircraft. Boyington arrived just before CASU-11 in March 1943. Lindbergh spent time on Guadalcanal just before CASU-11 started shipping out for home in June 1944. There's no evidence that either had personal interaction with CASU-11, but both became very well-known after the war. Many Sailors who served on Guadalcanal talked about being on the island simultaneously with these two gentlemen. Little, specifically, remembered the guys in CASU-11 talking about Boyington as stories migrated from one camp site to the next.

On 11 March 1943, Boyington arrived as the Executive Officer of Marine Fighter Squadron 122 (VMF-122) on Guadalcanal. When CASU-11 arrived on 18 March 1943 he was actively involved in flying defensive patrols over the island. By 19 April 1943 he was the Commanding Officer of VMF-122. At that time, he was not well known, and he was not happy flying only defensive missions. He wanted to chase and fight the Japanese over their home airfields. Finally, he became Commanding Officer of VMF-214 and moved north up the Solomons to the Russell Islands. On 16 September 1943, he was credited with shooting down five Japanese aircraft during one mission. A pilot became an ace when he has shot down five enemy aircraft. Now as an "Ace," he quickly became known as "Pappy." His squadron continued moving up the island chain and finally, could easily reach Rabaul. On 17 December 1943, Boyington led the first fighter sweep over Rabaul. Allied bomber pilots celebrated this event because they could now expect continuous fighter protection from takeoff to landing at their home base.

Pappy Boyington in cockpit receiving stickers for aircraft he shot down. Marine Photo.

By 3 January 1944, Boyington had shot down 26 enemy aircraft and was flying out of an airfield at Vella Lavella. "Japanese fighters were waiting and though he shot down two more Zeros to bring his score to twenty-eight, his wingman was shot down, and Boyington himself finally had to jump from his burning Corsair. Wounded, he floated in the water until he was taken prisoner by the crew of a Japanese submarine. Sent to a prison camp in Japan, Boyington waited out the last 20 months of the war."[49] He was later awarded the Congressional Medal of Honor.

Tues. 20th No raid. Not even alert. Going to pay us once a month. So no pay day. Wed. 21st Alert but no raid. Have 12 to 4 tonight. Lick had bad luck today. Anti-tank shell exploded. Blew off left hand, right thumb, and stomach wound. Not expected to live. Native observer hurt. Finally died, severed artery. Thurs. 22nd Two alerts no raid. Lick still alive. From the diary of Harry Hays, April 1943.

Lick was injured while handling anti-tank gun ammunition causing him to suffer critical, life threatening injuries. A native who was watching the ongoing activity was hit by the explosion and died from a severed artery. The local Navy infirmary kept Lick alive during the night but he was still in critical condition the following morning.

Fri. 23rd Nothing new. Peaceful night. Sent Lick to Espiritu. Sat. 24th All quiet. Sun. 25th Easter. Went to Mass. It is now 1:15P.M. and we're on alert. Must be a raid somewhere. Everyone by their fox holes. P-40s taking off. We're hoping for the best. All clear at 1:40. Good. From the diary of Harry Hays, April 1943.

Catholic Church Services. From John McAteer.

The Curtiss P-40 fighter was the Army Air Corps aircraft designed to protect bombers. It had the long range capability necessary to stay with the bombers during an entire mission from takeoff to landing.

Curtiss P-40 Army Air Corp airplane. Army Photo.

Mon. 26th Quiet night. One of the fellows caught a wild parrot. It was sitting on a jeep and he flashed his light on it and blinded it then picked it up. It bit his finger. Sure is pretty. Bright red, green and black. Tues. 27th Alert but no raid. Am back in metal shop. Laid a small concrete floor in the hopes will have a shower. Wed. 28th Quiet. Poured rain. Watch fired 4 shots at "spook." Bang, Halt! Thurs. 29th 2 alerts, no raid. Make 1st class beginning tomorrow. From the diary of Harry Hays, April 1943.

Guadalcanal has 181 species of birds that call the island home or visit on a regular basis. Nine of the species are parrots or parrot-related Lorikeets or Lorys. It's impossible to determine which species bit the finger of Hays's friend; however, it is likely the bird was beautiful and quickly gone. Sailors were given rifles and posted along the edge of the jungle. They shot at any unknown or strange noise.

Often the nighttime noise maker was a land crab. They were everywhere, and big ones could get as large as a foot across the back. As reported by Lane in his book, *Guadalcanal Marine,* "Those ugly, ravaging plunderers hid by day and roamed at night. Anything stored or left standing without surveillance was not safe from those scavengers. U.S. mailbags and anything else that had already been opened became their special domain. Packages from home were ripped open, shredded, and devoured by the land crabs. What they didn't finish, the ants carried away. We were only able to

81

get rid of the land crabs by stepping on them or pounding them with shovels. After they were squashed, we had to bury them in a hurry. Dead land crabs in Guadalcanal's hot humid environment emitted a nauseating stench."[50]

Following the Allied capture of Guadalcanal, there was a pause in military operations in the Pacific theater. This was a result of unexpected difficulties in the Mediterranean campaign, requiring more resources for use against the Germans than previously planned. Additionally, during April 1943, the USS Enterprise and the crippled Saratoga were the only available Pacific carriers. The new Essex-class carrier would not be available until August 1943. This required Nimitz to slow offensive operations at sea and focus on air operations. This seemingly negative pause had a positive outcome, as it gave American manufacturers time to greatly increase their output of carriers, battleships and attack transport ships.

"The best in new naval construction [carriers and battleships] was deliberately hoarded at Pearl Harbor for the eventual drive up the center [the Western Pacific from Guadalcanal to Mainland Japan.], which was to begin as soon as the build-up permitted. The Japanese, noting this growing concentration of power, also held back their capital ships [battleships]."[51] As it turned out, this limited, aviation-only, hand was given to just the right player, Admiral Halsey, whose attitude was to "Keep pushing the Japs around."[52]

Late April 1943 the four squadrons of Carrier Air Group Eleven (CAG-11) arrived over Guadalcanal. One squadron was composed of 35 F4F-4 Wildcats and they were assigned to operate out of Lunga Field (Fighter 1). Next were two squadrons of SBD-3 Dauntless Dive Bombers, totaling 35 more aircraft. They were directed to fly out of Henderson Field. And finally, the fourth squadron was 18 TBF-1 Avengers, who were also tasked to fly out of Henderson Field.

Another 88 aircraft for Cmdr. Schlossbach and the men of CASU-11 to maintain and repair. Harry Hays mentioned the arrival of these airplanes in his 4 May 1943 diary entry.

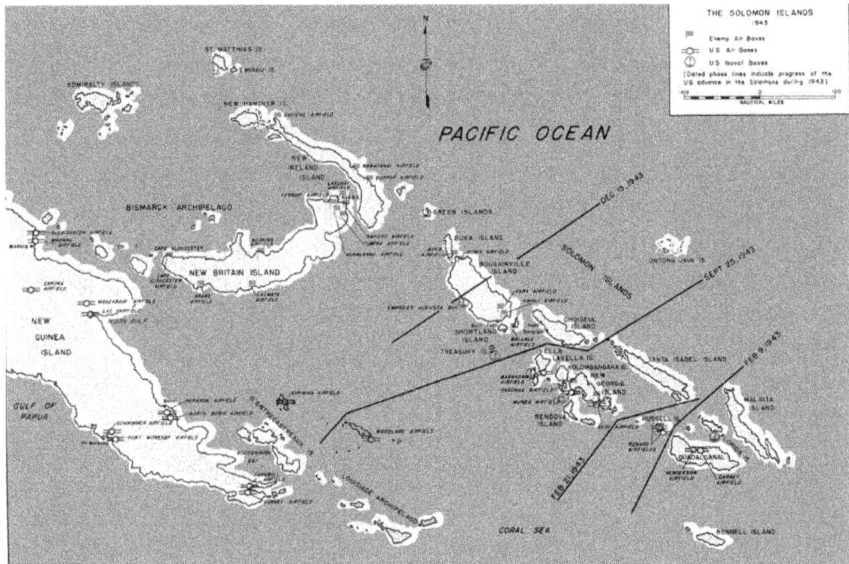

Capture dates as Marines moved up the Solomons. Marine Corps Graphic.

From March 1943 through June 1944, the entire time CASU-11 was in the South Pacific, "there were no carrier engagements at all anywhere in the Pacific. [It was a war carried on exclusively by land-based aircraft. And] yet in terms of sorties flown, aircraft shot down, bombs dropped - indeed by any reasonable measure - the air war in the South Pacific was fought, won, and lost by these land-based air forces"[53] In the absence of numerous Allied aircraft carriers, the only location in the Pacific where you could find an adequate number of islands, spaced close enough to support WWII era aircraft flight ranges, and with enough flat land to support runways, was in the Southwest Pacific. The Allies had no alternative strategy except to take advantage of these island air bases, later to become affectionately known as "coral carriers" or "stationary flattops." The Japanese also wanted these island airports, and that is the primary reason Carrier Aircraft Service Units were created. These air bases became the only strategic object of importance to both sides, and wherever you have aircraft, you must have aircraft mechanics.

In spite of only having a few ships and limited air forces, which were scattered across three islands, New Caledonia, Espiritu Santo and Guadalcanal, Halsey went to work. "During ... April 1943 the Americans

83

had the advantage in quality aloft and sometimes in numbers too. South Pacific Command (SOPAC) had about 316 planes of all types on Guadalcanal and these could be augmented by a pool of over 200 more aircraft at Espiritu Santo and New Caledonia."[54] On top of better numbers, the Allied air force was improving in quality with every plane that arrived during the multi-month pause. "They were now higher powered, more heavily armed and armored, and with longer range."[55] Against this limited force, the Japanese launched their best aircraft and most experienced pilots. At this point neither side knew who would be the ultimate victor of this phase of the Pacific war. Twenty months later, after Rabaul was no longer viable, the scoreboard below showed why the Allies succeeded:

Allies	Japanese
- One joint all service commander coordinating all air missions.	- Separate Navy and Army leadership with little coordination.
- Self-sealing fuel tanks.	- No fuel tank sealing.
- Seat armor.	- No seat armor.
- Island hopping bypassed Japanese strong points and gained airfields.	- Slow loss of airfields and ground forces left personnel trapped on isolated islands.
- Organized pilot rescue, seaplanes ships and submarines.	- No rescue, many very experienced pilots left to perish by drowning or to starve on remote islands.

These differences resulted in a campaign that had enormous significance. The Allies accrued five advantages from this campaign. "(1) It enabled MacArthur to break through the Bismarcks barrier of enemy air and sea power and advance toward the Philippines, (2) it bypassed and put out of the war more than 125,000 Japanese troops, (3) it reduced Japanese air power to the point where it was no longer a serious threat, (4) it forced the Japanese to withdraw their carriers from the Pacific, and (5) it gained for the United

States time to provide ships, weapons and trained manpower for a swift advance across the Central Pacific."[56]

In addition, Japan lost nearly 1,000 navy aircraft trying to recapture Guadalcanal and about 1,500 more defending the Upper Solomons and Rabaul. As a result of these accomplishments, for the rest of the war, the Allies put well trained and experienced pilots up against minimally trained and inexperienced Japanese youth. And there was CASU-11, right in the middle, running Henderson Field and doing their best to repair, maintain, and arm aircraft as fast as possible in order to get them up in the air for the next challenge.

May 1943

CINCPAC *Operations in Pacific Ocean Areas* report for May 1943: The month of May, in the South Pacific, witnessed an increase in our [Allied] aerial bombing of enemy positions in the Solomons and a diminution of enemy aerial activity. The major engagement occurred 13 May, at which time the enemy made its sole air effort of any magnitude. At 1100, 13 May, a coast-watcher reported a flight of enemy planes enroute to the Guadalcanal area. An hour later, Henderson Field radar made contact with this group bearing 107 degrees. The attackers turned out to be 'Zekes.' The reason for their appearance without bombers is not clear. Probably it was a reconnaissance mission looking for TF 18 [Navy Carrier Battlegroup] which had visited them in a combined bombardment and mining operation in the Munda-Vila-Kula Gulf [New Georgia Islands] area the previous night. The largest force of Allied fighters (102) to be scrambled during the war in the Solomons was sent up in defense. At 1300, 26 Zeke's were met by 14 VF (Fighters) 23 miles southwest of Cape Esperance [the northwest tip of Guadalcanal.]. After first contact, a total of about 75% of our air-borne fighters made contact with the enemy, shooting down 17 of them, (8 at the point of original contact). We lost 5 planes, and 3 pilots.

Zeke was the Allied nickname for the Mitsubishi A6M Zero, the Japanese premier, long-range, fighter aircraft. The Zero enjoyed an outstanding reputation as a fighter when attacking Allied aircraft.

Sat. 1ˢᵗ Alert no raid. Sun. 2ⁿᵈ No alert, but spooky night. Terrific rain yesterday. Bunk and mattress all wet. Most of clothes, too. Mon. 3ʳᵈ Quiet. Tues. 4ᵗʰ Quiet. I guess those two air groups that came in scared Bogey. Hope so. One group has 90 planes. Wed. 5ᵗʰ Pay day. $162.00. Drew and sent to my baby. Quiet. Had 12 to 4. Thurs. 6ᵗʰ Quiet. Fri. 7ᵗʰ Quiet. Sat. 8ᵗʰ Never heard it rain harder in my life. Buckets. More spooks. (Ha). Lick died on way home. Sending Daugherty away. Yellow Jaundice. Task force supposed to be heading this way. Hope its scuttlebutt. Sun. 9ᵗʰ Mother's Day. No Mass on account of rain and wind. Many tents with water over decks. Tops of trees blown off. All O.K. here so far. Mud, mud, mud! Mon. 10ᵗʰ Quiet. Tues. 11ᵗʰ Quiet. Sun Helmets issued. Also two shots. Wed. 12ᵗʰ Quiet. From the diary of Harry Hays, May 1943.

A "bogey" was an aircraft whose identity was unknown. Aviation Ordnanceman Second Class Jess William Lick died on 23 April 1943. He died from his injuries during the plane ride or shortly after arriving at the Navy hospital, Espiritu Santo. He was the first man lost from CASU-11 during the war.

Mud and More Mud in Camp. From John McAteer.

Sun Helmets were actually "pith helmets" made from woven dried pith. They were lightweight and excellent head gear for the tropics.

Thurs. 13th Raid this afternoon. Listened at radio shack. Angels – altitude. Knucklehead – Recon – Here. "Can I come in, cannon all shot up." Saw plane dive in ocean. Pilot jumped. One F4U got 4 out of 5 Zeros. Saw two F4U's come in shot up. "Wheels locked. Can't land. Running out of gas." From the diary of Harry Hays, May 1943.

Corsair F4U Fighter Aircraft. Marine Corps Photo.

Code name (Knucklehead) was for Russell Island, so the Japanese would not know what American pilots were talking about during their radio conversations.

Japanese Zero Fighter Aircraft. Navy Photo.

Fri. 14th Raid but no A-A fire. Night fighters got two of them. Sat. 15th Quiet. Sun. 16th Alert but no raid. Went to Communion. Mon. 17th Quiet. Rained all night. Tues. 18th 4 hour raid. Mostly Tulagi but we got some, too. Wed. 19th Boy, what a night! 9 alerts and 4 raids. From 8 P.M. until 5:00 A.M. Had 1 hour sleep. Saw night fighters get Jap bomber. Will never forget it. Everybody yelled and whistled. 9 fellows killed in N.O.B. Bomb dropped so close it looked like day. I was first in Foxhole. Tevis' on my back and Barker on Tevis's back. Griswold in sick bay. Feel dopey today. From the diary of Harry Hays, May 1943.

"A-A" was anti-aircraft gunfire from Army guns surrounding Henderson Field. "N.O.B" was the Naval Operating Base. It consisted of the areas around the CASU-11 campsite and housed a variety of Navy organizations, such as the expansive Navy supply area, other tent locations for troops, the hospital and massive amounts of crated supplies.

Thurs. 20th 2 hours alert but no raid, thank goodness. Dive Bombers with Fighters Escorts followed by Heavy Bombers coming over. Intercepted at Russells Island so didn't complete mission. Fri. 21st Quiet. Had 4 to 8. Sat. 22nd Crocodile shot in river we've been bathing in. About 7' long. Also 3 snakes killed in foxholes that was underwater. Ye Gods! Sun. 23rd 3 alerts, no raids. Last one we were up before "Condition Red." Sound of exploding depth charges at a sub woke us. Went to Mass. Father Heber Missionary officiated. 3 weeks back from Bougainville. Heard another crocodile screaming river during Mass. Mon. 24th Raid. About 1 hour. Bombs and all the trimmings. Tues. 25th Quiet. Wed. 26th Quiet. MacArthur, Knox and Halsey supposed to be here on Island. From the diary of Harry Hays, May 1943.

Bombed Radio Station near Henderson Field. Navy Photo.

Bombed B-24 Bomber Parked at Henderson Field. Navy Photo.

Guadalcanal Crocodile. Navy Photo.

Crocodiles could and did scream, and they could sound very human-like. Hays's mention of the island's visitors must have been a rumor. MacArthur never visited Guadalcanal during the war. Secretary of the Navy Knox visited in January 1943. Halsey was already there as he was the admiral in charge of the Southwestern Pacific Command located on Guadalcanal.

Left to Right, Secretary Knox, hands on hips, then Nimitz, without helmet, and Halsey with cigarette in mouth. Navy Photo.

Thurs. 27th Quiet. Rained all night. As we are screening tent damn thing nearly fell in on us. Everybody up to bail water. Sat around by flashlight and ate sardines, crackers, apricots and cherries. From the diary of Harry Hays, May 1943.

Men were provided with rolls of window screening with a fine mesh capable of stopping mosquitos. However, figuring out how to employ it around a flapping canvas cloth tent tied down with ropes was another story.

O'Flynn also spoke about the food, "Well, everything was right there" he said. "Our tents were right there. Mess hall was right there, once they got it built. Seabees built the restaurant. When we first got there, it was open air. They made stoves out of 50-gallon drums. Everything cooked in the open there. You take your mess kit, and you go through the line, put your stuff in it and go off in a corner somewhere and eat it. We had corn beef hash, wieners, spam for seven months. And sea biscuits, you had to have hard good teeth to eat the sea biscuits. The food was all canned rations, spam, wieners. We had this other combination of a thing. It was - I can't get the words for it - anyway it was roast beef, corn beef hash, out of the can, rolled up. They made stew out of it; they cooked two or three meals out of it. That's what we had for seven months. We got pretty tired of it. I still don't eat wieners anymore. Spam - I eat a little spam."

From O'Flynn: "One morning during the almost daily Japanese bombing runs, a couple of bombs fell into the supply dump in the middle of crates of food stuffs. The trail from the CASU-11 tent area to Henderson Field was nearby. After the 'all clear' was sounded, men resumed walking the trail to and from work where they found the trail littered with hundreds of cans of spam. Undamaged cans were picked up and hidden under practically every bunk in the area, while the damaged cans were pitched deeper into the jungle or used to bait rat traps." Little remembered that there were very few vegetables available. However, some of the Sailors would climb the palm trees and cut out the new shoots. He said they were very tasty.

The men of CASU-11 were opportunistic eaters. Little recalled having both a hankering for peaches and an opportunity to steal a number ten can of the fruit. He took the can, buried it, and began thinking about when he could eat it. A few evenings later, he retrieved the can. He ate the peaches while sitting on an empty beach at night enjoying the stars. He noticed what looked like fireworks on the horizon - it was nighttime action over Tulagi, 12 miles to the north. It was probably a Japanese destroyer firing rounds into the seaplane base. Having consumed all the peaches, Little went back to his tent to sleep. He woke up later with nighttime action of his own. He was violently sick to his stomach, and lost his peaches. Little never ate another canned peach again.

A lot of thievery went on. The supply area was piled high with boxes and crates loaded with food and sundries. Everybody performed after-dark raids on these supplies. Robert Leckie, a Marine, reported some thieves were judged more sensible than others. "No frippery, no useless ornament or artifices of that artificial world back home, like electric shavers or gold rings or wallets, nothing but solid swag of the sort that was without price on our island, things like socks or T-shirts or bars of soap or boxes of crackers."[57] Leckie further shared that in one of his midnight raids into a guarded tent holding "highly pilferable materials," he first found boxes of cookies, which were very sought after. But then he found cigars, "If cookies were worth their weight in gold on Guadalcanal, then cigars were worth theirs in platinum. In value, cigars could only be surpassed by whiskey."[58] Like Little, Leckie had a craving for fruit. He found a gallon can of preserved apricots. The end result was similar to Little's reaction. The fruit tasted great going down. But then he wrote it made him "wonderfully, wonderfully sick to my stomach. I lay on my belly and felt the stretching pain and marveled: 'I'm sick. I ate too much. It's the most wonderful thing in the world - I ate too much!'"[59] The next picture shows an early Army warehouse. It looks very vulnerable to intrusion and the removal of pilferable items.

Army Warehouse near CASU-11. From David Wilson.

It wasn't just food and supplies that were taken, Little revealed, "One of the tricks we used to play: When we went down to Guadalcanal, we had two jeeps and one truck, and a couple of motorcycles we stole off the Marines. We used to detour where the creeks ran and made great big mud holes, and when the Army would come riding by, they would get stuck in the mud hole while we was sitting out in the jungle on solid ground with a vehicle that would winch their vehicle out of the mud hole. We quickly took it back to our outfit, gave it a coat of paint and a Navy number, and then we had an extra vehicle. And when I left, I think we had around 150 vehicles that, ah, all of it belonged to the government, nobody brought any home, they still stayed. Maybe they are buried down there somewhere, don't know if they ever went back to gather them up or not?" The answer to this last statement was that vast amounts of materials, at war's end, were just buried or dumped into the sea and forgotten.

Next Jeep to be caught in Little's mud trap? Marine Corps Photo.

Jeep labeled CASU 11. From John McAteer.

The previous picture is of three men in a CASU-11 jeep, left to right, they are Lt McAteer, Unknown, and SK Chief Smith. Could this be a jeep Little swiped from the Army? We will never know. Two more pictures of CASU-11 men with jeeps follow and their poses seems to reflect feelings of jeep ownership. These men were all enlisted, and they were not usually issued jeeps. Had Little been at it again?

CASU-11 men, Nelson, Dickey, and Parker. From David Wilson.

CASU-11 Radio, Radar and Electrical personnel with their jeep. From Durward West.

Fri. 28th Quiet. Put in charge of general work at shop. Tent still not completed. Looks a little bad from outside. Sun. 30th Memorial Day. Went to

Mass. Beautiful Sermon. About 5 waves of B24's went out with P40 and F4U escorts. Went right over our heads while the sermon was going on, and during rest of Mass. Looked like hen with a brood of chicks. Mon. 31ˢᵗ Quiet. Rained all night. More rain today. Tent nearly finished. From the diary of Harry Hays, May 1943.

B-24 Liberator Army Air Corp Bomber. Army Photo.

O'Flynn mentioned that, "You had people [coming and going] getting transferred, going to school, and everything else all through the year and a half we were out there." For instance, Lt.j.g. John McAteer, arrived on Guadalcanal in late May 1943 to become the new Disbursing, and then Supply Officer for CASU-11. Many of the photos included within these pages are from his personal camera. The details of his trip to Guadalcanal are provided below because, while his trip was not unique, it shows that officers were often routed through myriad paths and means.

On 21 April 1943, Ensign McAteer was received onboard USS Munargo (AP 20) moored at Pier 45, Berth B, North Point, San Francisco, California. He had orders to CASU-11 on Guadalcanal.

The Munargo got underway 29 April 1943 at 0814 in the morning. The ship's commanding officer, Cmdr. McGee, had orders, issued by Commandant 12th Naval District, to proceed to Noumea, New Caledonia. On 17 May 1943, after an uneventful transit the Munargo moored starboard to Dock # 8, Point Doiambo, Noumea, New Caledonia. Four days later USS Munargo proceeded to Ebon Atoll Marshall Islands.

On 22 May 1943, the Munargo departed from Ebon Atoll bound for Espiritu Santo Island, New Hebrides. Two days later, on 24 May 1943, Munargo anchored in Segond Channel Espiritu Santo, New Hebrides. It is unknown what ship McAteer rode to finally complete his trip to Guadalcanal. Appendix E provides additional details about the Munargo.

Another man who arrived after the main part of CASU-11 was Aviation Metalsmith Frank Parker. On 21 July 1943 he checked in with CASU-11 as part of a small group on men arriving on Guadalcanal. His son, John Parker, shared some words about his father's ship ride in a 17 December 2017 letter written to the author,

> None of the men had ever before seen an ocean. Not sleeping well for most of the voyage, one day my father found a rare quiet spot on the ship to nap during the daytime. Although he was wearing a skivvy shirt, the ship had entered the tropics and dad was seriously sunburned. In fact, at sick bay they had to cut his shirt off while pouring vinegar over him. He never forgot that day.

> Since the ship was also carrying equipment and supplies, the decks were stacked, especially with trucks one above another. On one day that had especially high seas, my father and a buddy climbed up to the cab of the highest stacked truck to avoid the water spray. He said they were shocked when the bow of the ship scooped under a high wave and threw water completely over the truck with them inside.

In the spring of 1943, Guadalcanal was, "No longer known in code as 'Cactus,' the Guadalcanal base was now referred to in dispatches as 'Mainyard.' And a main yard it was indeed, with a thousand mainland factories and a hundred other bases funneling materiel into the supply dumps around Lunga Point … supply dumps everywhere."[60] Material came in so fast and so furious because the offloading ships, when at anchor, felt vulnerable to air attack. They offloaded as quickly as they could to return to relative safety away from port. So the beaches and near-beach areas were jammed with thousands of crates and boxes. They arrived too fast to sort and label. CASU-11 would regularly send men to the supply dump with hammers to bust open crates looking for aircraft parts.

O'Flynn noted that the supply dump generally had needed parts, "If you had to overhaul an engine or something, you could get the parts there. It may take a couple of days to get them rounded up, but they could get them. Well, all the stuff was new though, so we had very little calamities with it. We never lost any parts to getting bombed or anything." O'Flynn also reported that, "you have to understand that if you needed something you went out and got it, whatever you had to do. If you could do it and keep peace in the family that was all well and good. If you didn't, it might get a little ragged."

The cause of these problems was threefold. First, the supply system was capable of delivering materials faster than the receiving island could handle them. Second, there were not enough personnel on the island to unload and process all of this incoming material. Third, Guadalcanal had limited personnel to load cargo and the available inter-island shipping was too small to handle the volumes of freight arriving daily. Bigger ships were necessary to complete the follow-on delivery up the Solomons. However, most of the large ships were already headed home to bring another load across the Pacific.

Both Navy and Army vessels carried their respective required materials. Decisions made by the two services did not always synchronize with the needs on the receiving end. Consequently, personnel found ways to cope by meeting the unit's needs without being in accordance with proper procedures. Admiral Halsey stated in a major investigation on this topic, that "practically every man on Guadalcanal - Army, Navy and Marines alike - participated in the improper procurement of provisions and supplies, and the sales and trading that resulted from the securing of such items."[61] He went on to state there was no "Black Market" because there was no money to support it and no third-party population to sell the military goods to for profit. It was his opinion that, for the most part, it was a Sailor, Soldier, or Marine just trying to get the parts needed to fix something broken or to stop one's own stomach from growling because of hunger.

June 1943

CINCPAC *Operations in Pacific Ocean Areas* report for June 1943: The month of June in the South Pacific was characterized by continued training, improvement in facilities, increased air activity, and the commencement of the New Georgia Operation, which resulted in the occupation of Wickham Anchorage, Viru Harbor, Rendova Island and the ultimate reduction of Munda.

In the early part of June the enemy made two strong fighter plane sweeps, apparently without bombers, the purpose of which was not wholly clear unless they believed that such sweeps resulted in a ratio of attrition favorable to them, thus reducing our interceptor strength and leaving us more vulnerable to bomber attacks later. The numbers they employed seem excessive for armed reconnaissance alone. In the first sweep, on 7 June, by 50 Japanese VFs, 110 of our VFs were scrambled to intercept them over the Russell Islands, 24 of the enemy being shot down to a loss of 7 of our own. In the second sweep on 12 June by 40 enemy VF's, 118 of ours took off to intercept, with a resulting loss of 23 to 6 in our favor.

On 16 June, the largest enemy force since 7 April attacked our shipping in the Guadalcanal area. Enemy forces consisted of at least 60 VB (Bomber), screened by a like number of fighters. 104 U. S. fighters were scrambled in defense, and 74 made contact with the enemy. There were numerous U.S. ships in the transport areas off Lunga Point and in Tulagi. The attack lasted from 1315 to 1513. ComSoPac credits his VFs and ships' A/A with the destruction of 107 enemy planes. Six of our VFs were lost, two pilots being recovered.

Tues. 1ˢᵗ More rain. One big puddle. Quiet. Wed. 2ⁿᵈ Quiet. Rained all night. Thurs. 3ʳᵈ Quiet. More rain. Fri. 4ᵗʰ Quiet. More rain. Sure is slushy. Sat. 5ᵗʰ Alert, but no raid. More rain. Sun. 6ᵗʰ Quiet. More rain. Went to Mass. Someone hunting crocodiles. Heard gun fire and crocodile scream. Mon. 7ᵗʰ Quiet. More rain. First day off. Handed in course. Poured rain. Had 1 hour raid. In foxhole all the time. Some day off! From the diary of Harry Hays, June 1943.

It rains as much as 160 inches per year on Guadalcanal. Hays' reference to "course handed in" meant the correspondence courses Sailors took to qualify for promotions.

Seabees in Guadalcanal flood. Navy Photo.

Here is the official Army description of this period on Guadalcanal, from *The Army Air Forces in World War II, Vol IV, The Pacific: Guadalcanal to Saipan August 1942 to July 1944*, Section II: Target Rabaul, Chapter 7: The Central Solomons.

During the latter half of May, Japanese retaliation for the aerial assaults upon their bases was sporadic and weak, but the general lull ended with the coming of June. For one thing, the new Russells strip was ready [for US use], permitting more frequent and effective fighter sweeps over the Buin area. For another, the deadly SBD's [Navy Bombers] now carried 50-gallon auxiliary fuel tanks which permitted them to reach Buin from their Solomons bases. These developments aroused the enemy. He shifted more of his air strength back to the Solomons, and on 7 June he began to use it, for he could scarcely be unaware of the preparations under way on Guadalcanal. On the Rabaul airdromes air strength again rose to a high level, showing 225 aircraft. In the Rabaul harbor area lay nearly fifty ships, and the enemy's search planes were very active, increasing both in number and range.

On 7 June, the Japanese began a [week long] air assault against Guadalcanal which surpassed anything yet attempted. The pattern was unchanged: large forces of fighters would escort the dive bombers against shipping off Guadalcanal. U.S. fighters out of Russell would open the action which might extend right down to Guadalcanal itself. These attempts were costly to the attackers, twenty-three Zeros going down on 7 June, four of them to New Zealand P-40's, now in its first major action. Moreover, every single pilot was recovered from the nine Allied planes lost by U.S. Fighter Command, which set the day's exchange at 23 to 9, a rather costly business to the enemy. On 12 June came the second heavy thrust, this one costing the Japanese thirty-one planes for a loss to Fighter Command of six planes and two pilots; next a lull for three days, and then by the 16th the enemy was ready. Search planes counted 245 planes at Rabaul and found the other fields jammed with aircraft.

Coast watchers' reports indicated something more than a normal fighter sweep was under way, and they were right. Ships lay off Tulagi and Guadalcanal. To attack them the Japanese converged on their targets with an estimated 120 aircraft; and Fighter Command, with ample forewarning, had in the air a total of 104 defending fighters of all services. The resulting clash constituted the greatest single Allied aerial victory of the Solomons campaign, as the air over Savo Island, Tulagi, Cape Esperance, and Koli Point was filled with enemy and Allied planes whirling about in dogfights, many of them amid the flak of ground and ship gunners. By 1403 the enemy was in full retreat, leaving behind so many down aircraft that the defenders were hesitant to believe their own results. An accurate estimate was impossible; much duplication occurred, but Fighter Command listed no less than 49 Zeros and 32 dive bombers as victims of its planes, with the F4F's claiming 30 kills and the P-40's 25 - all this at a cost of 6 Allied planes, AA gunners claimed seventeen more for a total of 98 out of the original estimated force of 120. Although the enemy escaped with no more than a bare handful of planes, he left behind a reminder of his visit. Very real damage occurred on the ground and on three vessels off Guadalcanal, two of which had to be beached off Lunga. Altogether losses afloat and ashore reached twenty-five killed, twenty-nine wounded, and twenty-two missing, but they could have been far more disastrous.[62]

Tues. 8th Quiet. Wed. 9th Quiet. Thurs. 10th Quiet. Fri. 11th 2 Alerts, no raid. Moon is coming up again. Almost took a nice trip to Cape Esperance and Savo [island five miles north of Cape Esperance.] *today, but Army guy didn't show up. Got up at 4:45, too. Sat. 12th Had 2 hour raid today, but not over us, so we just stood by. We are supposed to have knocked down 11 and lost 3 out of Air Group 11. That doesn't include the Army.* From the diary of Harry Hays, June 1943.

Noting Hays' comment on Air Group 11 losses, personnel losses were always very hard to take. Pappy Boyington, noted in his autobiography, *BAA BAA Black Sheep,* that, "the majority of pilots in the war were not shot down by the enemy; they were killed in operational accidents in taking off from the fields, in getting lost in the fog, and so forth, not by enemy fire."[63] Other causes of pilot losses included running out of fuel, colliding into each other, shooting each other by mistake, shot down by own anti-aircraft guns, struck by falling ordnance from friendly aircraft above, and pilot errors.

Everybody knew combat flying was hours and hours of dull routine laced with a few seconds of stark terror. However, the wait for your "pilot" now overdue, grated upon everyone. This was acutely felt by the "enlisted plane captain" who treated the plane as if it was his own property and the pilot was his brother. The unexplained loss of a popular pilot or a skilled enlisted man rippled through the tents of CASU-11. Was the loss caused by a mistake while working on that plane's engine or flight controls? When other pilots gave eye-witness reports explaining what had happened to the plane, then there was closure and reluctant acceptance. It was the mysterious disappearances that lingered in the mind and hurt the most. So many men were lost without explanation; no one saw the incident. A pilot flew into a cloud and never reappeared, did he hit a mountain? Was he shot down? An aerial fight began with tens of Allied aircraft against tens of Japanese planes and after 10 to 15 minutes of intense chaos the planes part. Then, each group headed to their respective bases and counted noses as they returned. Where was Harry and George? Did anyone see what happened to them? Too often no one replied, no one saw what happened to their partners. The fight had been too intense to catch sight of a plane going down. Another pair of pilots was lost without explanation.

Bergerud states, "Yet despite the tremendous obstacles posed by geography, climate, microbes everywhere, poor facilities, and dangerous weather, two great air forces engaged in a two-year aerial slugfest over the South Pacific. To fight this war, both sides required complex and powerful aircraft that appeared in a remarkably short period of technological advance in military aviation."[64]

Sun. 13th Quiet. Went to Communion. Went to Army movies on Fighter Strip. 2 features, news reel, 3 comedies. Had 2 hour alert while we were there. 3 waves of Jap Bombers, but they never came in. Sure glad, as the strip is no place to be in a raid. Mon. 14th Day off. Washed clothes. Went for ride with Griswald all over Island. Went swimming in Koli River. Swell. Tues. 15th 1 hour raid. Caught us napping again. I was only one dressed when bombs fell. About 6 of them. How did London stand it? Launched the boat. 2 P-38 auxiliary gas tanks. Works swell. Have 12 to 4 Jr O.D tonight. From the diary of Harry Hays, June 1943.

For recreational purposes the Sailors made boats from empty P-38 Drop Tanks. These were easily obtained from the nearby Army Fighter strip, Kukum Field (Fighter 2). They were fashioned into a floating raft by using some wood or metal cross braces. The pictures below shows the tanks and the boat.

P-38 Drop Tank. Army Photo.

Drop Tank Turned Into Boat. Army Photo.

Wed. 16th Had 1 hour raid last night. Tulagi got most of it. It is now about 2:30. The sirens just sounded the red alert. Supposed to be about 8 miles out. I'm all sweaty from running to our tent from the shop. Boy, what a fight! I haven't any idea what time it is, but so far we are supposed to have shot down 31 Zeros and 19 Dive Bombers. They sank at least one of our cargo ships, and left two burning. Don't know how many planes we lost. Dog Fights and strafing right over our fox hole. Bullets are being found in our division and the 2nd, too. Saw Jap bombers peel off and dive on one of our ships. It's burning now and has been beached to keep it from sinking. They're towing another one in from the middle of the bay. Smoke is pouring out of her stern. Change that to 31 Zeros and 29 dive bombers. Latest report. Air Group 11 got 28 themselves. Final dope 80 to 6. From the diary of Harry Hays, June 1943.

Official Navy records do not show a ship sunk during this 16 June 1943 Japanese attack off Guadalcanal, only noteworthy ship damage was to the two ships detailed below.

While attempting to defend itself, the USS Celeno (AK-76), an Attack Cargo ship, took two direct hits and multiple near misses. The crew was credited with aiding in downing at least three enemy planes and damaging several others. In spite of serious damage, the ship managed to safely beach itself. Fifteen of her crew were killed and 19 wounded in the attack. Celeno was towed to Espiritu Santo for repairs followed by five months in the San Francisco Shipyard. The ship returned to the Solomons January 1944.

The burning ship that was towed to shore and beached off Lunga Point was the Landing Ship Tank 340 (LST-340). The air raid severely damaged the ship with several near misses and one direct hit by a 300-pound bomb on the main deck. One crew member and nine Army passengers died in this attack. LST-340 was towed to Espiritu Santo, repaired enough to safely travel to San Francisco, and after a full overhaul it returned to action ten months later.

Thurs. 17th ½ alert, but no raid. It's about 9:30 A.M now and we just got the report that bogey was 30 miles out. No alert sounded as yet. Condition Green. They were our bombers with no I.F.F. on. Fri. 18th 2 hour raid in harbor and Tulagi in Bay, what fireworks. It's now about 2:00 P.M. Another Condition Red. "They didn't come over, they're afraid of us." "I'm running low on gas." "Aren't we all?" Sat. 19th Quiet. Sun. 20th Quiet. Went to Mass. It is now 12:00 and the alert just sounded. I haven't been to chow, either. 12:45 All Clear. Drove Jeep today. First car I have driven since we left. Mon. 21st 2 Alerts, but no raid. Day off. Looked for grass skirts but no go!! Tues. 22nd Alert, but no raid. Wed. 23rd Same. Thurs. 24th Poured rain again. Alert, no raid. Fri. 25th Alert, no raid. Sure do miss home tonight. Sat. 26th Alert, no raid. Sun. 27th Quiet. Mon. 28th Quiet. We have to have gas masks ready at all times from now on. Japs gassed New Guinea. Tues. 29th Quiet. Went to work in fabric shop. Like it better. Wed. 30th Quiet. Have 8 to 12. Back in welding shop. From the diary of Harry Hays, June 1943.

Out For a Drive on Guadalcanal. Marine Corps Photo.

I.F.F. meant Identification Friend or Foe. It was an electronic device installed on U.S. aircraft that responded to radar pulses with a unique signal that would identify an airplane as friendly (Allied) or as an enemy (Japanese). Hays has quoted some of the radio chatter between pilots that he overheard while standing near the radio tent. Between 15 through 26 June, five Condition Red Alerts occurred, all in the early morning hours.

The official historical records do not substantiate Hays' claim of the Japanese using gas on New Guinea. However, during post war interrogation, Japanese officers revealed the extensive use of chemical and biological weapons from 1931 through to the end of the war in August 1945 in Manchuria, China. Secret bases were established for experimenting upon animals and humans, and 1,000 mustard or lewisite gas attacks were conducted against the Chinese.

During late June into late July VC-28 (A Composite Squadron with both fighter and bomber aircraft) was based at Henderson Field. For that period they stated in their war diary, "an extremely efficient CASU [11] was established at Henderson Field, and it did an excellent maintenance job on

106

the planes. Most work relating to gunnery, communications and material condition of aircraft was handled by the CASU."[65] Additionally, they noted Quonset huts were now available, vice the mud floored tents they had lived in during their March visit, and the food was better.

The increasing flow of Army aviation personnel onto Henderson Field lead to many new faces. Schlossbach was now designated as the Commanding Officer of Henderson and Lunga Fields along with CASU-11. He issued the pass below to Hays. This certified that Hays had proper authority to approach airplanes parked at Henderson Field. These security measures were required to protect the aircraft and stop the theft of tools and test equipment on the flight line where all of CASU-11's aircraft were parked.

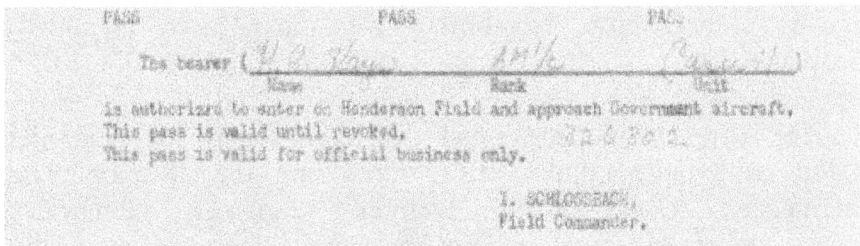

Henderson Field Pass. From Harry Hays.

July 1943

CINCPAC *Operations in Pacific Ocean Areas* report for July 1943: The final offensive to capture the Japanese base of Munda, on Southwest New Georgia immediately surrounding the Munda Point airfield, opened on 30 June. It was the first combined land, sea, and air effort undertaken by our forces in the Solomons since the capture of Guadalcanal. The participation of the air force was on a greater scale than at any previous period during the Solomons Campaign and the problems involved, largely because of distance, were difficult of solution.

The campaign lasted 37 days and included the protection of our convoys, the covering of landings, and numerous missions, many of them on a large scale for this area, in support of our troops ashore, as well as attacks on the nearby Japanese bases of Vila, Balbale and Kihili.

During this period our air force reported the destruction in combat of the following Japanese planes:

Fighters		259
Twin-engine bombers		60
Dive bombers		23
Float planes		16
	Total	358

Our combat losses totaled 71 fighters and 22 bombers of all types during the same period. On the Munda target area alone more than 900 tons of bombs were dropped in approximately 30 strikes. There were approximately 100 other strikes during the period, 60 of them against land targets and 40 against shipping. During the period, 35 fighter pilots were recovered out of 81 forced down in combat or in operational accidents during combat flights. Many of these were picked up by 'Dumbos' (PBY5 seaplanes).

The Allies and Japanese had a different regard for their pilots. The Allies deployed maximum effort in rescuing downed pilots. They used high-speed rescue boats, "Dumbo" seaplanes, submarines, and coastwatchers to recover and put back into action every pilot they could. The Japanese did very little to rescue their pilots, and this was a mistake. To the Japanese, pilots were expendable and, as a result, many experienced pilots perished needlessly. This was one of the most critical mistakes made by the Japanese during World War II.

Thurs. 1st Quiet. We're supposed to have landed on Munda yesterday. Sure is a lot of activity in the air. The Chief started making the guys who come late to work stay late. I don't blame him. Fri. 2nd Quiet. Wind storm came up all of a sudden. Blew tops of trees off. Sure fierce. Sat. 3rd Quiet. Heard today we're having tough time on Munda.75 out of 100 lost in 12 minutes. Had to withdraw and reform on account of heavy casualties. 3 hours late in reaching objective. Searchlights have been going in circles all night to lead planes in. From the diary of Harry Hays, July 1943.

On 30 June 1943, over 6,000 Soldiers and Marines landed on Rendova, five miles from Munda Field, to start the push to take the airfield from the

Japanese. Starting out with great optimism, things rapidly changed for these Allied troops. The Japanese had 4,500 troops placed in difficult jungle conditions with a well-entrenched perimeter. Combat was so bad on Munda Field that a directive went out tasking all available aircraft, pilots, and aircrew to be ready on short notice to fly missions up to Munda Field. CASU-11 was assigned a number of test pilots, radiomen and gunners who tested all repaired airplanes to ensure the planes were combat ready. The July 1943 Battle for Munda Field required the participation of all CASU-11 ready aircraft and aircrewmen. CASU-11 aviators joined in flying Munda Field missions with the Navy SBD Squadron VC-28, the same squadron that was currently stationed with CASU-11 at Henderson Field.

July 1943 page from Aviators Flight Log. From Durward West.

This page from the Flight Log Book of Aviation Radioman Second Class Durward West, was graciously provided by Allison West, his daughter. West flew seven strike missions during July as radioman for various pilots in Dauntless Dive Bomber (SBD) airplanes. He accumulated 28 hours of combat flight time.

On 25 July 1943 West participated in a 174-plane air raid, the largest Allied strike to date in the Pacific. Two days later West received the following commendation signed by W.A. Shryock, Commanding Officer, Composite Squadron 28 and I. Schlossbach, Commanding Officer, CASU-11.

This man is commended for outstanding services rendered to the pilots of VC-28. During the period from 1 July 1943 to 26 July 1943 this man voluntarily serving as radio-gunner, participated in numerous and extended bombing missions on enemy territory. These missions were accomplished in the face of heavy anti-aircraft and enemy fighter opposition. The enthusiastic cooperation and exceptional ability shown by this man materially assisted the pilots with whom they flew in inflicting serious damage on the enemy. His cool bravery and high devotion to duty is in keeping with the highest traditions of the United States Naval Service.

West standing third from left Radio and Electronics CASU-11
Guadalcanal. From Durward West.

On 5 August 1943, after 37 days of intense fighting, Munda Field was finally taken. This battle was so difficult that it had taken almost 50,000 additional men versus the 6,000 who landed on the first day. Munda Field was one more stepping stone that would bring our aircraft within striking range of the Japanese fortress at Rabaul.

On 25 July 1943, Admiral Marc Mitscher departed Guadalcanal after passing command of Air Solomons to Army Air Force General Twining. Mitscher had arrived in February to command all aircraft in the Solomon Islands. Taylor in his book, *The Magnificent Mitscher* reported that as Commander Air Solomons Mitscher had led his forces in destroying, "the air threat to the Solomons. His box score was 340 Japanese fighters and 132 Japanese bombers destroyed, 17 ships sunk and 8 damaged. He'd spilled 2,083 tons of bombs on the enemy, but the single tally of which he was proudest was 131, representing that number of United States and New Zealand pilots who had been snatched from alien waters and beaches by his rescue planes."[66] CASU-11 maintained the fighters, the bombers, and the rescue aircraft that helped Mitscher defend Guadalcanal and inflict great pain upon the Japanese airmen. And what did Guadalcanal give Admiral Mitscher as a parting gift? He was bitten by a mosquito with malaria. Already in poor health and possibly recovering from a heart attack, he arrived home weighing 115 pounds.

Sun. 4ᵗʰ of July. Quiet. Went to Mass. More searchlights. Mon. 5ᵗʰ Quiet. Rained all night and all day. Tues. 6ᵗʰ Alert, no raid. Wed. 7ᵗʰ Quiet. Thurs. 8ᵗʰ Have 4 to 8. Quiet. Griswold in Sick Bay with Malaria again. Sending Lingbeck out to graft skin on his knees. Sores keep breaking out on an old scar. Changed back to C.A.S.U. #11. Lord only knows what that means. Sure miss home tonight. From the diary of Harry Hays, July 1943.

After malaria the most common Guadalcanal affliction was jungle rot. Specifically it was a fungus on the skin that resulted from being continuously wet. The term jungle rot or the crud also included a variety of other skin ailments, such as, athlete's foot, rashes, impetigo and scabies which could happen just about anywhere on your body. Primary locations were the feet, thighs, face and scalp. "The primary treatment was a solution of potassium permanganate, which was irritating to the skin and stained it purple, but was effective at killing the fungus."[67] Most CASU-11 tents had a bottle available with a brush attached to the lid for application. There was always one or two men using it.

Fri. 9th Alert, no raid. 5 months today. Sat. 10th Quiet. Went after poles for welding shop. I never hope to see such jungles again. I believe those Tarzan pictures now. Sun. 11th Alert, no raid. Went to Mass. Mon. 12th 3 Alerts, no raid. Heard Artie Shaw. Plenty good. Tues. 13th Alert, no raid. Wed 14th Quiet. Rain again. Thurs. 15th Quiet. Fri. 16th Quiet. Sat. 17th 3 Alerts, one raid. Just like old times. Fox hole and everything. Had Jap in lights, but no hit by AA. Just heard that 72 men were killed in Carney Field last night. Delayed action bombs. Didn't explode until alert was over. Sun. 18th 3 Alerts, 2 raids. Real thing, again. Saw A20A fire 2 bursts of tracers at Jap bomber. Don't know whether   he got it or not. Boy, I'm sleepy. Heard that 11 came over and we got 3. Went to Communion. Will probably be another bad night, as the moon is still almost full. From the diary of Harry Hays, July 1943.

In the period from 5 to 12 July, there was at least one Condition Red Alert each day. A20A is shorthand for 20 caliber anti-aircraft gun. These were the guns placed around U.S. airfields to defend them from enemy aircraft.

Anti-Aircraft Gun Emplacement near Henderson Field. Army photo.
Mon. 19th One alert, no raid. Went to Hill #27 today. Also Native Village. Some trip. We really saw fox holes. Bought grass skirt. Saw bones, shoes, shells, helmet, etc. From the diary of Harry Hays, July 1943.

In early January 1943, Hill #27 was captured by U.S. forces, just days before Guadalcanal was declared enemy free. The Japanese lost an estimated 500 soldiers on Hill 27, which they called The Battle of the Mountain of Blood. On the four nights from 16 to 19 July, there were 10 air raids within the hearing range of Hays on Guadalcanal.

Searching for souvenirs or war relics from the surrounding battlefields filled the spare time of many Sailors and Soldiers. Native carvings and grass skirts were sold by the locals, military machinists would turn brass gun shell casings into collectables, and just about everyone could collect sea shells for lei-making. Wives and girlfriends especially appreciated souvenir shell leis and grass skirts because they represented the beauty of the tropical islands and not the ugliness of war.

Across all of the Solomon Islands the most wanted object was the grass skirt. On Guadalcanal, "Solomon Islanders were 'selling grass skirts to G.I.'s at fabulous prices.' In order to get more sales they added color. Around Munda Field they dyed [the grass skirts] yellow using atebrin (Malaria medicine) tablets."[68]

War relics, military objects collected from the battlefield, were also highly sought, sometimes too highly, by many men. Indeed, as soon as the fighting ended, the souvenir hunting began. Robert Leckie, recalled in his book *Helmet For My Pillow*, that "moving among them [the dead Japanese] were the souvenir hunters, picking their way delicately as though fearful of booby traps, while stripping the bodies of their possessions."[69]

Members of CASU-11 who were interested in the war relic industry found themselves on both the making and the receiving and distributing ends of the business. Marines passing through from various island battlefields often had stuff to sell, and pilots bouncing on Henderson Field for fuel and ammo often had ready cash and/or a bottle of whiskey ready for trade.

When the Army passed through Guadalcanal the Soldiers came with hundreds of cases of assorted candy bars. Bartering activity would begin as soon as two men were close enough to talk to each other. Hershey bars or Butterfingers were exchanged for a sword or flag. Often the flag was made by the Marines and to make it appear more authentic it had Japanese script copied from Japanese canned goods. The Marine Pioneers, according to Guadalcanal veteran Kerry Lane, in his *Marine Pioneers: The Unsung Heroes of World War II*, traded souvenirs to the Soldiers for "food, new socks, dungarees and even shoes."[70] The Marines also traded or sold souvenirs to the crews of ships that had brought the Soldiers to Guadalcanal.

Newspaper correspondent Ira Wolfert, in his 1943 classic *Battle for the Solomons* recalls the story of an American pilot in mid-October 1942 on Guadalcanal who told him of a conversation he had with a captured Japanese bomber pilot. He claimed to have been a graduate from Ohio State University. The Japanese said they were fighting for Togo and the Germans were fighting for Hitler, "but your Marines seem to be fighting for souvenirs!"[71]

Foster also mentioned, "Every day a Marine, Soldier or Seabee wandered up to the pilots with something to sell for cash or whiskey. Jap rifles cost from $25 to $100, or one to three quarts of whiskey. Japanese invasion currency or their national currency sold for three pieces per $1 for the bills with the larger dimensions and $1 for four pieces of the smaller-sized dimensions. The price of a Japanese flag consisting of a square piece of white silk with a large red round sun in the center, with various Japanese hieroglyphics (and perhaps a bloodstain or two), was from $50 up to $300, or three to six quarts of whiskey."[72]

Proud Souvenir Owner. Marine Corps Photo.

Bennett talked about this robust trade in her book, "When the Australian infantrymen on their six shillings daily wage saw the U.S. demand for valued Japanese swords soar to 50 pounds, [Word made it back to Australia and] 'one of our base workshops in Moresby started to manufacture swords, which by far were the most sought after articles. They did this in style with an assembly line and a weekly output target. The blades were made of car springs, beautifully shaped by the blacksmiths ... often better and more beautifully finished than the genuine article.' These blades were probably as good as the ordinary swords of the Japanese infantryman."[73]

Real or phony sword? Army Photo.

Bennett went on to note, "Seabee entrepreneurs would buy Japanese souvenirs from enlisted men and sell them to others 'at a profit, which is a common practice on the island.'" Attempts were made to regulate the market. "On Guadalcanal, trading in artifacts was confined to the 'highway' during daylight on Sundays."[74] The natives quickly figured out the schedule and soon a whole fleet of native canoes was regularly seen on Sunday mornings travelling toward Lunga to trade with the American military.

Gunnery shell casings were reworked into "inkwells, ashtrays, bottle openers, picture frames and rings."[75] These items certified an individual had served in combat even though, for most men, they had never served on the front lines. They just went down the road on a Sunday and bought something that looked interesting.

And, finally, the "short snorter dollar" was a unique souvenir of the war. This dollar bill was signed by close members of a group, tent mates, work mates, etc. The dollar bill pictured below is from Allison West who kindly shared this photo. It was signed by members of Durward West's work group.

Short snorter dollar of Durward West. Photo from Allison West.

Tues. 20th One hour raid. Wed. 21st Quiet. Thurs. 22nd Quiet. Had 4 to 8. Fri. 23rd Quiet. Winter moved in. Barker has ringworms all over his feet and 2 under one arm. Sat. 24th Quiet. Sun. 25th Quiet. Went to Mass. 154 planes supposed to have taken off for Munda this A.M. at 4:00 o'clock. Hope they settle the argument. Mon. 26th One Alert, no raid. Day off. Went for another ride. Nothing unusual. Tues. 27th Quiet. Wed. 28th Quiet. Had to muster at 5:55 A.M. in order to try to get better chow. Thurs. 29th Quiet. Fri. 30thQuiet. Poured rain again. Sat. 31st More rain. Quiet. From the diary of Harry Hays, July 1943.

Admiral Halsey and Nimitz knew that they would need more airplanes and ships to eliminate Rabaul as an active protagonist against Allied forces in the Southwest Pacific. Aircraft that were essential in the capture of Munda Field could now operate from that airfield and attack Rabaul. More airplanes were rolling off the assembly line and were ferried to Guadalcanal and Munda Field. With an increase in aircraft and the decrease flight time to Rabaul, the Allies increased their daily sorties. Pressure on Japanese forces centered on Rabaul was increasing daily.

August 1943

CINCPAC *Operations in Pacific Ocean Areas* report for August 1943: During August there was a continuation of the general progress in augmentation and training of all forces, in improvement of facilities ashore and in consolidation of newly acquired positions.

Air operations were of the same general nature as heretofore, and consisted of (1) searches, (2) A/S [Anti-submarine] patrols, (3) air coverage, (4) interceptions of enemy air efforts, and (5) offensive missions, including support for ground forces, bombing enemy installations and attacks on enemy shipping. Japanese air efforts were principally in the nature of harassment of our bases, and movements of men and supplies.

On 13 August an unknown number of Japanese bombers and torpedo aircraft made a rather skillful 'sneak' raid on Guadalcanal, sinking John Penn (APA) with an aerial torpedo. Some of the Japanese aircraft trailed behind a flight of U.S. heavy bombers returning to Guadalcanal from a strike in the north, these enemy planes proceeded to drop bombs and flares over Guadalcanal. Attention was diverted to this attack meanwhile, 8 torpedo planes, flying low and fast, came in over Florida Island to attack our shipping, dispersed throughout the approaches to Guadalcanal.

John Penn, received the brunt of the torpedo attack. At 2123 a torpedo hit the starboard side, ten minutes later abandon ship was ordered, and at 2150 the Penn sank stern first with 7 officers and 91 men killed and missing.

Sun. 1st Quiet. Went to Mass. Mon. 2nd Quiet. Day off. Washed clothes and went for walk along beach. Uneventful. Tues. 3rd Quiet. Wed. 4th. Quiet. Have $302.00 on books. Think I'll leave it there. Heard that starting Monday we get every other day off. Hope so, as there is very little work. From the diary of Harry Hays, August 1943.

On 1 August 1943, CASU-11 had 17 officers, 3 warrant officers and 602 enlisted men. Hays' report of little work was the result of Allied aircraft moving to airfields further up the Solomon Island chain. CASU-11 was told to stand down and only work every other day. This pause would not last very long. Nimitz was receiving reports on the production at the shipyards and aircraft factories back home, and he knew mid-September he would have a large surge of ship and aircraft deliveries.

Munda Field, New Georgia Island. Marine Corps Photo.

Thurs. 5th Quiet. Fri. 6th Quiet, but probably not for long. Moon is beginning to appear. Put in for $50.00 Bond every month. Sat. 7th Quiet. Sun. 8th Quiet. Went to Mass. Starting tomorrow every other day off. Mon. 9th Quiet. Tues. 10th Quiet. My day off. Don't know what to do. Did nothing. Wed. 11th Quiet. Raining. Draft going home. About 7 men out of the shop, but not us. Darn it!! Thurs. 12th 1 Alert no raid. Moon is out again. Fri. 13th Some date. Heard that 5 Japs were coming over last night, but our fighters got to them first. Have 4 to 9 in the morning. Sat. 14th 2 hour 35 minute raid last night. Bad one, too. Jap torpedo planes got one transport. Luckily it was empty. One 200 yards away was full, but they chose wrong one. The planes were right over the water and right over our heads. The tracers from the 50c were terrific. We got one of them. Bombers came over, also. Don't know how many, but we saw three, and heard two more. Also got one of them, but two planes won't make up for a transport. Thought we lost our packages from home, but found them in freight office. Sun. 15th 2 Raids. 1 for ½ hours and 1 for 51/2 hours. What a night! Saw our night fighter fire tracers at one bomber. Don't know whether he got him or not. 5 bombs dropped so near we could smell the powder, and dirt fell in our fox holes. We saw 9 Jap bombers, and no telling how many more there were. From the diary of Harry Hays, August 1943.

A 50-caliber machine gun was called a "50c." Fifty caliber was the diameter of the bullets, and in this case it was 0.510 inches. Most fighter aircraft had two or four of these machine guns mounted on their wings. Today these "packages from home" are called care packages and they usually contain cookies, candy and personal items military personnel are unable to find on base. Working every other day with daily bombs dropping around you must have been extremely depressing. Daily work took your mind off the exploding bombs and made you feel that your efforts were contributing to winning the war.

Mon. 16th Quiet, which sure surprised us as the moon was full. Maybe our bombers were keeping them busy. Oh yes! Went to Mass yesterday. Off Day today. Washed clothes, played catch, badminton, acey dewcey, etc. Tues. 17th One alert, no raid. From the diary of Harry Hays, August 1943.

Usually spelled "Acey Deucey," played with a board and dice, it was a form of Backgammon. Between 1 August and 18 August there were 10 Condition Reds set and highlighted by a big raid on Friday 13 August 1943. During this raid SS John Penn was bombed at anchorage and sunk. Fortunately the troops and cargo were already discharged.

Wed. 18th Two alerts, no raid. Heard today we might have to stay here 18 months. Heaven forbid! From the diary of Harry Hays, August 1943.

Every day the men asked the same question: When are we going home? According to O'Flynn, no one knew when CASU-11 was going home until late May 1944. On 30 June 1944 a large group of men of CASU-11 left Guadalcanal.

Thurs. 19th One alert, no raid. Fri. 20th One alert, no raid. Beer night. Ah! Sat. 21st One alert, no raid. Sun. 22nd Day off. Quiet. Didn't go to Mass, as we left about 8:30 to get shells. Mon. 23rd Quiet. Tues. 24th Quiet. Day off. Washed and walked along beach. Got sunburned. Wed. 25th Quiet. All bitten up. Mosquitoes must have been in tent last night. Thurs. 26th Quiet. Turned in Progress Course. Fri 27th. Quiet. Asked Mr. Ricktenwald about transfer to North Island. Sat. 28th Quiet. Sun. 29th Quiet. Went to Communion. Mon. 30th Quiet. Had 8 to 12. Tues. 31st Quiet. From the diary of Harry Hays, August 1943.

September 1943

CINCPAC *Operations in Pacific Ocean Areas* report for September 1943: In the South Pacific, the air effort increased from 2,409 sorties in August to 2,553 sorties in September. Approximately 90% of these were concentrated against [enemy] land targets divided approximately as follows: Bougainville [Island] 55%; Kolombangara [New Georgia Island] 34%; Choiseul [Island] 5%; and Ringi-Webster Cove barge center 5%. [These are the locations of Japanese airfields and harbors in the northern half of the Solomons.] During September several night air attacks were made on our positions at Guadalcanal and Munda Field. We received moderate damage.

Carriers are now present in the Pacific in sufficient numbers to begin damaging attacks all along the enemy's eastern defense perimeter. Until now we have been unable to carry the war to the enemy in the Central Pacific area except by submarine.

Wed. 1st Quiet. Thurs. 2nd Quiet. Fri. 3rd Quiet. Pay Day. No more days off. We're taking over Henderson Field. Sat. 4th Quiet. Sent $80.00 to my baby. Sun. 5th Quiet. Went to Mass. Mistake made. Ordinance gang were disarming Jap ammunition and burning the powder. Fire got away from them and ignited our dump. Shells were exploding from 2 o'clock in the afternoon until 2:30 last night. 90's block busters, etc. Boy, the time and money that was wasted. A number of men hurt by shrapnel, and 6 men killed. All planes moved from Fighter #2 to Fighter #1, and field secured. Could have been a field day for the Japs. Glad they didn't know it. From the diary of Harry Hays, September 1943.

While "quiet" generally meant there were no air attacks on a specific day, quiet did not mean there was no work being done. Everyday equipment was repaired that had been damaged the night before. All aircraft had to be inspected to determine if there was damage from the previous night's bombing. Equipment used for repairing aircraft - for example, gear for pulling engines, replacing tires or aiming guns - needed continuous attention. Mechanics always needed parts. These were found in the thousands of crates among the coconut palms around the Supply Center or found in the wrecks lying near the various airfields. Some mechanics kept hand drawn maps of selected wrecked aircraft. Often it was easier to send a Sailor to pull the part from a previously damaged airplane than to guess which box held the desired pieces.

Mon. 6th Quiet. Tues. 7th Quiet. Wed. 8th Quiet. Got 2 packages. Heard 30 piece Marine Band. Swell. Jimmy Cathcart with them. Thurs. 9th Quiet. Fri. 10th Quiet. Had 4 to 8 watch. Sat. 11th Quiet. Sure is monotonous. Wish I was home. Sun. 12th. I spoke too soon. 2 Alerts, no raid. One alert lasted 1 ½ hours, the next about ½ hour. Went to Mass. Mon. 13th Quiet. Tues. 14th 2 hour raid. No one in our tent heard "Condition Red." The Jap planes diving woke us up. They dropped about 6 or 7 bombs, and we all hit the decks when we heard them falling. No time to run for fox holes. Not a stitch of clothes on and Griswald all wrapped up in his blanket. Our 90's fired about 10 rounds and then were quiet. We dressed and spent next 2 hours by fox holes. Heard today that one bomb fell so close to a guy at the boat pool that it knocked him out of his sack, but no casualties. From the diary of Harry Hays, September 1943.

Guns for anti-aircraft defense were placed around Henderson Field including a few 90-caliber weapons shown in the next picture. These were the "90's" Hays reported in his diary.

Henderson Field Surrounded by A-A Guns. Army Photo.

Wed. 15th 1 hour raid. About 6 more bombs. Direct hit on SBD [Dauntless Dive Bomber]. 5 men killed and 7 badly hurt in Marine camp just across road from our shop. One bomb fell in front of radio shack. 19 holes in one tent. 2 tires on truck blown out. Truck full of holes. They're getting too close, uncomfortable. Thurs. 16th 1 ½ raid. Bombs dropped on Carney. Fri. 17th Alert, no raid. Sat. 18th Eleanor came in last night. Alert no raid. I think it was a show for her. About 6 P.M. Still light, lasted about 20 minutes, radio gave all the dope-"Bogey 5 miles to west. Bogey shot down" etc. From the diary of Harry Hays, September 1943.

Admiral Nimitz put everyone back to full-time work and First Lady Eleanor Roosevelt visited Guadalcanal. When Admiral Halsey first heard that she was coming to the South Pacific, he balked at her visiting the combat zone. However, he had a change of heart after her successful visit to Australia. "He had been hearing reports of her work and was impressed. Halsey decided that Eleanor had earned her trip to Guadalcanal.

Eleanor Roosevelt Talking with Troops. Marine Corps Photo.

When First Lady Eleanor Roosevelt arrived on Guadalcanal, the island was still being bombed by the Japanese. On Guadalcanal the men were not told of her coming, but the night before her arrival, they were told that they were not to walk around without wearing pants and shirts, as they often did.

The men on Guadalcanal were completely surprised to see the First Lady. One astonished Marine exclaimed, 'Gosh, there's Eleanor!' Her escorting general was disturbed by the familiarity, but Eleanor was amused. She made the rounds of hospitals, kitchens, a cemetery, workstations and tent dwellings of the men.[76]

She experienced an air raid and wore a steel helmet when she took shelter with the men. During her jeep tour of the facilities, the jeep inadvertently drove past an outdoor shower filled with naked Marine aviators. Although they were told she was visiting, they had just returned from a mission and were too tired and dirty to care - and neither did she!

Eleanor Roosevelt visiting Guadalcanal. Navy Photo.

This plane, pictured above, crashed near the 63rd Construction Battalion (Seabees), and after Eleanor left, the resulting wreckage was picked up, and used as war relic trading material.

She came finally to a chapel and a graveyard on Guadalcanal that touched her more than any place before. The chapel had been built of wood by the natives and given to the Americans. Its steep, thatch roof sloped almost to the ground. A thatch-covered cupola rose above and was topped by a cross. Eleanor reported that the local residents 'even made the altar and the altar vessels, carving them beautifully, decorating the church with symbols which have special meanings for them - fishes of various kinds which mean long life, eternity, etc.'

Outside in the cemetery, wooden crosses marked the graves of men killed in the fighting on Guadalcanal. Perhaps the simplicity of the markers most affected Eleanor. Mess kits and sometimes helmets hung on the crosses, and at the base of each marker friends had carved their prosaic tributes: 'He was a grand guy' or 'Best friend ever.' That was all, but here was war reduced to its plainest, the destruction of a young man, a great guy, a friend. In her diary Eleanor wrote, 'I said a prayer in my heart for the growth of the human spirit so that we might do away with force in settling disputes in the future.'[77]

Eleanor Roosevelt visiting Guadalcanal cemetery. Navy Photo.

125

Mrs. Roosevelt was not the first woman to visit Guadalcanal during the war. This honor was given to Flight Nurse Mae Olson, who was a member of the Air Evacuation Unit of the Army Air Corp. She stepped off an airplane while performing her duties associated with loading and transporting injured personnel. The men around the aircraft went wild. Word spread rapidly, resulting in men quickly gathering for just a glimpse of Mae. This occurred early June 1943, and by the end of that month, there were nurses working and living on Guadalcanal.

Flight nurse Lt. Mae Olson takes the name of a wounded American Soldier being placed aboard a C-47 for air evacuation from Guadalcanal in 1943. Air Force photo.

Sun. 19th 1 hour raid. More bombs. My birthday. 33 years old. Went to Mass. Heard that Munda Field is being bombed every 25 minutes. 60 planes in one raid yesterday. Hope we don't get sent there. Mon. 20th 1 hour raid over Carney. Fireworks were terrific. Tues. 21st 2 raids. First 1 ½ hours over Carney. Second 2 hours all over. Our night fighter P 38 shot down two that we saw and got another. We got 5 all told out of a possible 12. The rest were over Russell and Tulagi. Two men killed and four hurt by bomb dropped on Fighter #1. Weren't in their fox hole. Boy, it's a beautiful sight to see those planes fall. Just like a ball of fire. Wed. 22nd Quiet. Thurs. 23rd. Quiet. Fri. 24th Quiet. Sat. 25th Alert, no raid. Went to see cemetery today. Beautiful *church. All thatch, different designs, carved symbols, birds, fish. Sun. 26th Quiet. Went to Mass. Mon. 27th Quiet. Tues. 28th Quiet. Wed 29th Quiet. Thurs 30th Quiet.* From the diary of Harry Hays, September 1943.

Guadalcanal Chapel. From Robert Little.

"Munda [Field] is being bombed every 25 minutes." This was an important journal entry by Hays. While Hays regularly included "quiet" days in his diary, meaning no air raids, the pace of the work on aircraft increased daily. Unfortunately, he rarely talked about the work the men of CASU-11 were doing. During late September 1943, the Marines were having a tough time taking down the Japanese forces around Munda Field on New Georgia. The nonstop application of all available bombers was required, and this directly impacted the workload of CASU-11. Nearly every aircraft that returned to Henderson Field had a bullet hole that needed patching. As soon as possible, new SBD bombers were being unpacked and readied for flight. On 15 September an enemy raid destroyed one SBD and damaged four others and on 16 September a slightly damaged plane landed. On 20 September an SBD was slightly damaged by Japanese anti-aircraft gunfire and another one was lost as its engine cut out and it landed in the water. On 23 September an SBD was damaged when engine trouble required a soft landing. The plane flipped over, and, fortunately the crew was uninjured.

Wrecked aircraft Guadalcanal. Navy Photo.

In September the aircraft carriers Admiral Nimitz was told to expect began arriving. "The most important were the new fast *Essex*-class carriers. These were 28,000-ton ships bristling with antiaircraft guns ranging in size from 20mm to 5-inches and carrying the latest radar. Each carried almost a hundred aircraft. By October, Nimitz had ten of these fast carriers, as well as seven 11,000-ton *Independence*-class escort carriers, two new battleships, eight heavy and four light cruisers, and sixty-six destroyers."[78] Admiral Nimitz could now effectively support Admiral Halsey in attacking Rabaul and points north.

October 1943

CINCPAC *Operations in Pacific Ocean Areas* report for October 1943: During October the enemy was expelled from the lower Solomon Islands. Our aircraft flew 3,259 sorties in October, as compared to 2,553 sorties in September. We lost a total of 26 planes in combat, as compared to 49 in September. As before, over 90% of our effort was directed against enemy land targets, - principally in the Shortland and Buka-Bonis Areas, and on Choiseul Island.

That this effort was successful is proven by the fact that all Southern Bougainville enemy airfields were practically neutralized by the month's end, thereby removing one of the major hazards to the Treasury Island

Occupation, the Choiseul Diversion, and the projected operations against Bougainville. The enemy effort declined from 801 sorties in September to 495 in October, while their losses increased from 148 to 173, plus 16 destroyed on the ground.

On 10 October, about 0150, three enemy torpedo planes attacked merchantmen unloading off Guadalcanal. Two ships, SS John M. Couch and SS George Himes were hit.

Fri. 1ˢᵗ Quiet. Rained like heck. Sat. 2ⁿᵈ Quiet. More rain. Sun. 3ʳᵈ Quiet. More rain. Went to Mass. Mon. 4ᵗʰ Quiet. More rain. Tues. 5ᵗʰ Quiet. More rain. Pay day $77.00. Didn't draw. Wed. 6ᵗʰ Quiet. Regular storm. Poured rain. Thurs. 7ᵗʰ Quiet. Fri. 8ᵗʰ. Quiet. Start to work on Field Sunday. 24 on 24 off. Something new anyway. Sat. 9ᵗʰ Quiet. Had 4 to 8 last night. Sun. 10ᵗʰ Quiet. Couldn't go to Mass, as I was out on the field. Directed a PBY and DC3 in. Saw Jap prisoner. Also 2 nurses. Terrible! From the diary of Harry Hays, October 1943.

Japanese prisoners of war were held in a guarded compound in an unknown location near Lunga Point, Guadalcanal. The average October monthly rainfall was 4.20 inches. The airfield Hays was talking about was Henderson Field.

Mon. 11ᵗʰ 3 Alerts. No air raid, but sub got in and sank one of our ships. Don't know whether we got sub or not. Correction - it was no sub, but 3 planes flying at about 25 feet. Radar couldn't pick them up. And they sank two ships. Also had our I.F.F. signals. New order tonight is to shoot at any plane that comes within 40 miles of this Island. We have no planes at all in the air. From the diary of Harry Hays, October 1943.

"The Official Chronology of the U.S. Navy in World War II" reported that "Japanese planes attacked U.S. shipping off Koli Point, Guadalcanal, on Monday, 11 October. The Japanese torpedoed freighters *George H. Himes* and *John H. Couch*. Three men died on board *John H. Couch*, a merchant seaman, one Armed Guard Sailor and a CB stevedore. The *John H. Couch's* cargo included gasoline and diesel oil, which caught fire during the initial explosion. Firefighting efforts by [the crews of] two destroyer escorts proved as unsuccessful as the [Couch's] crew in putting out the blaze and the ship was abandoned by the 42 merchant seamen, 25 Armed Guards, 28 troop passengers and 99 stevedores. The gutted ship was towed by the tug *Pawnee* (AT-74) to a point two miles east of Koli Point, where the merchantman ship capsized. The *George H. Himes* was beached by the tug *Menominee (AT-73)* and suffered no casualties among the 41-man merchant complement, 27-man Armed Guard, and 20 CB stevedores. The ship's cargo of lumber, shells and bombs were saved."[79]

The "CB stevedores" loaded and unloaded ships and were part of special Construction Battalions established for this purpose. The letters "CB" were translated by Sailors to Seabees and became the common name for all the Navy Construction Battalions. Again I.F.F was Identification Friend or Foe. It was used to determine allied versus Japanese aircraft. Japanese engineers ascertained how I.F.F signals functioned and learned to spoof or mimic them. Some of their airplanes would send an identification signal matching those from Allied forces and slip into bombing range without coming under attack.

Tues. 12th One alert, no raid. Wed. 13th One alert, no raid. My day off. Also get 2 more days off. Not bad!! From the diary of Harry Hays, October 1943.

Early Tuesday morning, 12 October, Ship's Cook Juan Leal, Jr. was tasked with lighting a gasoline stove in the cooking area assigned to CASU-11. The stove exploded and threw burning gasoline over him. He suffered serious burns to his face, arms, hands, chest, thighs and back. He was taken to the nearby Mobile Hospital Number Eight where he died late on 16 October. Leal was buried in the Army, Navy and Marine Cemetery, Guadalcanal, Solomon Islands in grave 1, row 73. According to the *Brownsville Herald*, the newspaper of Brownsville, Texas, his remains were later moved to La Paloma Cemetery in La Paloma, Texas. This death and the death of Aviation Ordnanceman Lick, reported by Hays in his Diary on 8 May 1942, were the only known deaths of any CASU-11 personnel during their tour of duty on Guadalcanal.

Thurs. 14th One alert, no raid. We fired at a B-24. Glad we missed. Fri. 15th Quiet last night, but alert about 9:00 A.M. No raid. Saw Bill crying a little on the way home from the show last night. Sat. 16th Quiet. Sun. 17th Quiet. Went to Communion. Mon. 18th Quiet. Tues. 19th Quiet. Day off.

Nothing to do. Wed. 20th Quiet. Thurs. 21st Quiet. 3 days off starting today. Fri. 22nd Quiet. Sat. 23rd Quiet. Sun. 24th Quiet. Couldn't go to Mass on account of work on Field. Mon. 25th Quiet. Another bomb dump exploded last night. Had to help put out fire at one of the camps. From the diary of Harry Hays, October 1943.

Guadalcanal had one major ammo/bomb dump, a centralized storage area for weapons. Several smaller, ready-access dumps were located around each airfield. The major dump had weapons of all types while the smaller dumps were filled with weapons specific to the types of aircraft flying from the nearby airfield. There were even unidentified ammo dumps from the Japanese occupation. Occasionally mistakes were made, Astor recalled in his book, "Another time, the mosquito abatement guys burnt a grass field next to our Quonset hut and it turned out to be a Jap ammo dump...The dump started burning and shells got cooked off and kept flying out. It blew a large hole in our Quonset hut. We spent the whole next day in a fox hole." Astor continued, "one night, another tent not far from ours started to burn. Smoke and fire came out the top like a blow torch. We went down to see what it was. The entire tent burned up. The guys had a still and were making booze. Things just got out of hand."[80]

Tues. 26th Quiet. Have 12 to 4 tonight. Saw 2 nurses today. One not bad. Tipped my cap to her. Oh Boy!! DC taxied us today at over 40 miles per hour. F4U came in with wheels up and made belly landing. Brand new ship, too. Put my course in yesterday. Wed. 27th Quiet. Thurs. 28th Quiet. Fri. 29th Quiet. Saw 2 more nurses today. Sat. 30th Quiet. Sun. 31st Quiet. Went to Mass. About 60 Ordinance Men sent to Munda today. We might get it next. From the diary of Harry Hays, October 1943.

The Douglas C-47 (DC) passenger and cargo aircraft must have given Hays an exciting ride across the airfield. This was probably a ground test often performed after engine maintenance. The repaired airplane was intentionally taxied the length of the runway without reaching takeoff speed.

John Parker in a blog shared a story about a wheel's up landing from his dad, Aviation Metalsmith Henry Parker assigned to the metalsmith gang of CASU-11, "Repairing the shot up and damaged planes, my dad and his buddies became very close with the pilots. Once, after some repairs, a pilot asked my father if he wanted to go up on a test flight. Of course, my dad couldn't wait to go. Once in the air, everything checked out and they prepared to land. Unfortunately, they couldn't get the landing gear down. After a couple of attempts, no luck. Dad asked the pilot, 'What now?' The pilot responded, 'Well, we're going to climb up, then push the nose over and head for the ground. Then we'll pull all up fast and see if the gear comes out.' 'What if it doesn't?' Dad asked. The pilot smiled and asked, 'Do you know how to swim?'"[81]

Douglas C-47 aircraft, also known as Skytrain. Army Photo.

Ordnance men were trained in handling weapons and knowing how to load different types of aircraft with a variety of bombs, and they were now in high demand at Munda Field. On 5 August 1943 the Marines captured Munda Field, located on the southwestern tip of New Georgia Island, and by late October, this airfield was very busy as the Allied offensive continued its "island hopping" campaign north up the Solomon Island chain.

World War II aircraft made great demands on the pilot. Under the best of environmental and airfield conditions, there was always an element of risk when airplanes took off or landed. Aircraft with a tricycle wheeled undercarriage, pictured below, were vulnerable during maneuvers on the ground because of visibility difficulties. The pilots could only see to the sides of the airplane's nose, and they could not see straight ahead until enough speed was obtained to lift the tail. Many squadrons and CASUs would put a

Sailor or Marine on the wing to guide the pilot to the end of the runway or to his parking spot. Unfortunately, no hearing protection was ever used. Pictures of this are provided below and on this book's cover. This was not considered a safe procedure and eventually it became illegal.

Helping guide the plane. Marine Corps Photo.

The men of CASU-11 repaired aircraft on their "quiet" days. On 25 October 1943 an SBD was damaged by a bomb dropping from an SBD above it. A very lucky pilot brought the plane back to Henderson Field. On 26 October five SBDs were damaged by enemy anti-aircraft gunfire and safely returned to base. And on 28 October six SBDs were hit by Japanese gunfire with more bullet holes and internal aircraft damage.

If a plane was damaged during flight beyond its ability to continue flying, then the pilot had to make a choice, parachute into the ocean or into the jungle. Navy pilots always preferred the water.

Navy aviators were taught how to make water landings and exit a bobbing aircraft. They were equipped to go into the water with rafts, flares, some drinking water and rations. And they knew there were high-speed rescue boats, submarines, seaplanes, and ships pre-positioned and ready to come to their aid. On the other hand, the jungle would swallow an aviator into its greenness. Your parachute would tangle high in the trees and leave you many feet off the ground with the initial difficulty of descending to the ground. Once on the jungle floor, densely overgrown terrain filled with steep ridges and sometimes invisible valleys and ditches awaited. The jungle restricted pilots from seeing very far, and it was hard to find distinguishing landmarks. With thorny shrubbery and the jungle canopy holding in the heat and humidity walking was extremely difficult. Finally there was the almost torrential rainfall every day that soaked the wandering pilots. There was never a moment of dryness while taking that long walk.

Once home base was in sight, the landing on the runway was never easy. Causes of airfield crashes included: an out of fuel or damaged engine over airfield resulting in a "no power landing" or a "dead stick landing," poor alignment with runway during landing attempt, wheels that would not go down or just one wheel went down, damaged flight controls from bullets holes or pilot injury. Pilots also had poor visibility due to dust or "heavy rain or because the nose of aircraft restricted visibility on the ground. As a result, crashed remains of Japanese and Allied aircraft littered the space around every island airfield.

In his book, Foster provides an example of what can happen on a short, narrow, combat air strip on a jungle island.

One day a bunch of mechanics were out sitting on a small hill waiting on the return of the morning strike.

Two planes approached at 1,000 feet. Suddenly the wingman broke away from his leader and lowered his landing gear. 'That boy is in trouble - he's coming straight in!' When he saw he could reach the runway even if his engine failed - he popped his landing flaps to slow himself down.

'Good Gawd! Look what's coming in at the other end of the strip!' Another pilot dropped his landing flaps at the other end of the field. The two pilots were landing head on! Neither of them seemed to see the other.

Both planes were now skimming along just above the deck in the final stage of landing and both were landing long. At last the first plane's engine roared into life, its nose rose into the sky, the left wing dipped violently, missing the ground by only three feet or so as the pilot made an emergency left turn to evade the oncoming plane.

I turned my eyes to watch the other plane land. He had no sooner rolled the length of the runway and turned toward the taxi strip when his engine stopped.[82]

Later the mechanics learned the other plane made it out over the ocean, ran out of gas and dropped his plane successfully onto the water. This was quickly followed by a rescue boat pick up. Just another day at the jungle airport.

Wrecked aircraft Guadalcanal. Navy Photo.

Wrecked aircraft Guadalcanal. Marine Corps Photo.

O'Flynn maintained the cranes "that were used if we had a crash on the field somewhere. [We] had the cranes to clear the area. Took up the pieces of the planes." Little also shared, in a resume written well after the war, that he had the "job of picking up and salvaging crashed aircraft," and "loading and unloading [aircraft] carriers of airplanes." I asked O'Flynn if he remembered Little, his answer was no. However, I am sure they interacted with each other as Little checked out the gear from the motor pool where O'Flynn worked. This gear was necessary to pull another wrecked airplane out of the jungle for parts.

November 1943

CINCPAC *Operations in Pacific Ocean Areas* report for November 1943: During November, our aircraft in the South Pacific flew 4,481 sorties, as compared to 3,259 in October. We lost 69 planes in combat, as compared to 26 in October. About 82% of these sorties were against land targets, 13% were directed at shipping targets (including warships at Rabaul), and the remaining 5% were interceptions.

At the beginning of the month, our air operations had rendered enemy bases in the Shortland Area and in south and central Bougainville inoperative. Only on Buka Island and at Bonis, on the northern tip of Bougainville were Japanese airfields operational. These received carrier based attacks at the beginning of the month and were rendered unserviceable. There was only one small harassing raid at Guadalcanal.

Although the landing at Cape Torokina, Bougainville caused enemy air activity on a greater scale than any seen since the July attack on Rendova [New Georgia], this reaction was short-lived. The Japanese were evidently limited in replacements. In spite of tactics generally directed toward conserving air strength, enemy air opposition faded more rapidly than in July, and all air effort had ceased by 20 November except for small scale night harassing.

During November, aircraft carriers operated in the Solomons Area, supporting the amphibious landings on Bougainville. Strikes at enemy fields in the Buka-Bonis region at the moment of the initial landing were followed by heavy attacks on Japanese units afloat in the harbor of Rabaul.

Mon. 1ˢᵗ Quiet. Tues. 2ⁿᵈ Quiet. Wed. 3ʳᵈ Quiet. Thurs. 4ᵗʰ Quiet. Fri. 5ᵗʰ Quiet. Pay day. Didn't draw. Have XX on the books! Sat. 6ᵗʰ Quiet. Saw nurse in blouse and slacks and one in dress yesterday. She was sitting down and I saw both of her knees. First I've seen since we left. Haven't the slightest idea what she looked like. Sun. 7ᵗʰ Quiet. Couldn't go to Mass, as I had to be on airfield. Mon. 8ᵗʰ Quiet. Tues. 9ᵗʰ Quiet. Wed. 10ᵗʰ. Quiet. Thurs. 11ᵗʰ Quiet. Fri. 12ᵗʰ Quiet. It is now 2:30 in the afternoon and the Alert has just sounded. Bogey 17 miles out. 3:00 P.M. All clear. Sat. 13ᵗʰ. Quiet. Vega Ventura crashed on takeoff. Manned the fire truck, but no fire. From the diary of Harry Hays, November 1943.

Lockheed Vega Corp. Ventura (PV-1) Patrol Bomber Aircraft. Navy Photo.

The thirteen times that Hays noted quiet days in this November 1943 diary entry are worthy of explanation. Generally, it can be interpreted that Hays used the word "quiet" throughout his diary to mean no enemy air raids. However, from his first diary entries onward one should not take a quiet entry as meaning no work! There were always many airplanes to maintain and airfield work to be completed. By this time the Japanese effort to recapture Guadalcanal had cost them 1,000 naval aircraft. This loss also meant numerous Japanese pilots were lost, taking with them significant war-fighting experience. No wonder the Japanese raids noticeably declined.

Meanwhile, on the Allied side, many additional aircraft were arriving everyday along with more pilots. Experienced pilots trained new pilots, and the overall air-fighting capability improved in quantity and quality. New airfields were established under the Nimitz "island hopping" campaign. Each new airfield resulted in getting closer to Rabaul with an increasing number of Allied aircraft attacking Rabaul each day. "By the time Rabaul was eliminated in the spring of 1944, the Japanese had lost an additional 1,500 airplanes trying to prevent our movement north up the Solomons."[83]

This "XX" means Harry left some quantity of money on account with the Disbursing Officer.

Sun. 14th Quiet. Went to Mass. Have 12 to 4. Notice posted for 8 men to change to C.A.S.U. 15 Bougainville. If no volunteers, names to be drawn out of a hat. If I ever had any luck, I hope it sticks by me now. Mon. 15th Quiet. 6 volunteers. Drawing for 2 more to be held tomorrow. Tues. 16th Quiet. Applegate and Peterson. Boy, that was worse than any Bank Night I ever attended. Wed. 17th Quiet. Thurs 18th Quiet. Fri 19th Quiet. Sat. 20th Quiet. Sun. 21st Quiet. No Mass. Duty at Field. Mon. 22nd Quiet. Went up the coast again. Saw beached Jap Transports. Swam out to one and went aboard. Boy, what a sight! Plates buckled, big holes in her sides, all burnt out. From the diary of Harry Hays, November 1943.

In early December CASU-15 was scheduled to arrive on Guadalcanal and then later in the month continue on to Vella LaVella. Vella LaVella is an island in the New Georgia Island chain. On 15 August Allied infantry landed on Vella LaVella and by late October had complete control. A runway was surveyed, cleared and laid down during the ongoing combat. The first aircraft landed on 24 September. The men of CASU-11 were hoping to remain on Guadalcanal rather than going to Vella LaVella with CASU-15. Guadalcanal was receiving considerably less attention from the Japanese than those new airfields further up the island chain and, consequently, closer to Rabaul. A bank night was a recreational gambling night for all hands. Typically, poker, craps, and roulette were available. These bank nights were somewhat controlled by the officers and had low gambling limits.

Hays swam to a damaged Japanese transport ship. The plates he referenced were on the hull of the ship. The decks and sides of most ships were constructed by welding large steel rectangular plates together. This shipbuilding method was used prior to and during World War II by most shipbuilding nations.

Tues. 23rd Quiet. Wed. 24th Quiet. I have an infected finger from scratching a bite. Also have jock itch from playing volleyball. Thurs. 25th Quiet. DC came in today with flat tire. Circled for over 2 hours to use up gas. Made perfect landing. Wheelright turned jeep over and broke his leg. Winter wrecked water tank and cat. Very good Thanksgiving dinner. Fri. 26th Quiet. Took hop today. 7:00 A.M. until about 5:15 P.M. Koli Field-Russells-Segi-Munda-Ondonga-Villa La Villa-Munsla-Villa La Villa-Ondonga-Munda-Segi-Russells-Koli Field-Henderson. 14 takeoffs and landings. What a trip! The pilot was swell!! I have a very bad sun-burned lip. Hope it gets better. Another ammunition dump on fire today. It is after 7:00 P.M. and the whole sky over it is lighted by the explosion. I can't understand how those things happen. Sat. 27th Quiet. Sun. 28th Quiet. Went to Communion. Mon. 29th Quiet. Tues. 30th Quiet. Rain. From the diary of Harry Hays, November 1943.

Hell's Point Ammunition Fire. Navy photo.

CASU-11 had a fire truck and with a quickly assembled crew, it headed to the ammo dump fire. Aviation Metalsmith First Class David Wilson was on this fire truck and the following words, over Capt. H.A. Rochester's signature, were placed in his service record:

22 December 1943: During the Hell's Point Ammunition fire on 26 November 1943, at 0130, this man aboard an emergency fire truck, called from Lunga Field worked continuously to extinguish this fire until 0600

of that day. Explosions continued during most of this period, making work extremely hazardous. Also, much valuable and irreplaceable property was saved by this man's act.[84]

Wilson's selfless and automatic reaction in racing to fight this explosive fire was heroic by any measure.

The Hell's Point Ammunition Dump was established as a major Ammunition Dump for the South West Pacific area in late 1942 through early 1943. Located about a mile east of the northeast end of Henderson Airfield this roughly one square mile area became a major storage area for all types of U.S. ammunition. Captured Japanese ordnance found by the arriving Marines and ammo plowed up as Seabees constructed camp sites, runways and other structures were also stored there.

The ammo dump's location was close to the beach where ammo ships were unloaded and close to the aircraft at the airfield, which needed ammunition. However, there were a few problems with that location. As the weeks of 1943 rapidly went by, the war moved farther north, away from Guadalcanal, and less ordnance was pulled from the Hell's Point supply. Meanwhile, more ships were bringing ammo to the Southwest Pacific. The piles of ammo grew at the dump. With time and rain, the lowest levels of munitions began to sink into the soft soil. At war's end uncountable tons of ammunition remained in the Hell's Point ammo dump, much of it buried.

During the wet season, large amounts of grass grew rapidly over the fertile Guadalcanal plain. The dry season followed, and the grass quickly dried and became a potential fire threat. While the rules required the close mowing of the grass, this was rarely done. The brown grass became fuel for fires.

In November 1943, the fires began in multiple locations. A grass fire ignited the ammunition stored at Hell's Point causing a huge explosion. The fire lasted for three days.[85] Massive explosions resulted in Hell's Point being so badly contaminated with damaged ordnance that everyone living or working on Guadalcanal was restricted from entering this section of the island.

On a sad note after the war was over, "At Henderson's airfield it was found that the U.S. forces had bulldozed munitions into the ground. They (Australian Bomb Disposal Unit) excavated almost 20,000 projectiles, and still more remained."[86] The Hell's Point dump covered at least 100 acres and held about 15,000 tons of munitions. It was marked with "Off Limits" signs and to this day it is still closed.

With the Solomon Islands under the control of the Allies, hundreds of aircraft ready and loaded with ordnance, and increasing numbers of Allied ships in the Southwest Pacific, Admirals Halsey and Nimitz were ready to attack Rabaul with vigor.

Rabaul Harbor. Marine Corps Photo.

By early 1942, the Japanese military had turned Rabaul into a fortress with numerous anti-aircraft batteries and 100,000 Japanese troops. The volcanic island offered high, mountainous terrain around a large harbor. These geologic features were conducive to protecting aircraft and ships. Except for the harbor opening to the southeast the Japanese had installed numerous gun mounts on the mountain peaks which almost completely encircled Simpson Harbor. These gun mounts provided "overlapping zones of fire" that enabled adjacent guns to mutually support each other. The cross fire eliminated any uncovered ground and greatly reduced any chance of the enemy getting through. These gun mounts, in combination with the mountains, made Rabaul one of the most difficult

WWII targets in the Pacific. Both General MacArthur and Admiral Nimitz knew that the take-down of Rabaul was going to be necessary before commencing their two campaigns towards the Japan mainland.

The Scout Bomber Douglas (SBD) Monthly Sortie numbers detailed below shows what Admiral Halsey and Admiral Nimitz set in motion. In military aviation, a sortie is a combat mission of an individual aircraft, from takeoff to landing. For example, one mission involving six aircraft would tally six sorties. Comparing November and December sorties over Bougainville with the other months there was a noticeable upward trend in sorties culminating in the maximum effort into Rabaul from March through April 1944. The earlier outlier was the month of July when the Marine invasion of Segi Point, New Georgia to take Munda Airfield was supported by 1,124 SBD sorties.

The sortie information depicted below was extracted from the "War Diaries" of various South Pacific aviation commands, including Commander Air, South Pacific, Commander Air, Solomons, and Commander Air, Northern Solomons. The timeline for this sortie data covers the period that CASU-11 was assigned to Henderson Field, Guadalcanal, March 1943 through July 1944.

SBD Monthly Sorties 1943 - 1944

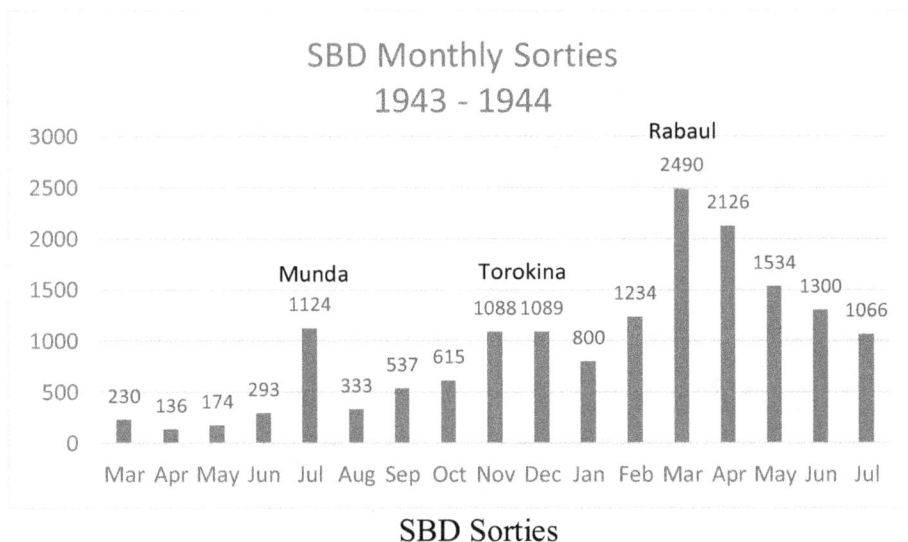

SBD Sorties

There was a total 15,422 sorties during this period with all sorties originating from Solomon Island airfields. One needs to imagine a sequentially growing number of aircraft runways located further, and further north up the Solomon Islands. First it was Henderson Field, Guadalcanal, secured 8 February 1943, and then Banika Field, Russell Island, opened 21 February 1943 – each assisting the Marines in taking the next airfield up the island chain with the ultimate goal of eventually ending Fortress Rabaul's capability to harm Allied forces. These two initial airfields were followed by Munda Field, New Georgia, opened on 5 August 1943, and then Vella Field, Vella LaVella with flight operations commencing 9 October 1943. And finally Torokina Airfield, Bougainville with initial flights 10 December 1943.

The goal was to shorten the flight time to Rabaul, pour massive numbers of aircraft out of the skies onto this Japanese fortress, and end all Japanese offensive capability in the area. March 1944 marked the end of major U.S. military operations against Rabaul. Three peaks are noteworthy; the first peak was the July 1943 targeting of Munda Field, the next peak was during November and December 1943 with 1088 and 1089 raids against Torokina Field, Bougainville, and the third, and final peak, was the massive effort March 1944 directed at closing Rabaul for good.

SBDs Departing Rabaul after Bombing Run. Marine Corps Photo.

It was June 1943 when CASU-11 began participating in the engagement of Rabaul, the deadly SBDs were finally equipped with 50 gallon auxiliary fuel tanks which enabled a flight to travel from Guadalcanal to Rabaul and back. The next two images are Pilots Reference Strip Charts which were carried by all pilots during World War II when flying in the Solomons. The first covers from San Cristobal, southeast of Guadalcanal, to Buka Island, which is northwest of Bougainville. Each ring is 50 nautical miles. Therefore, from Henderson Field to Torokina is about 350 nautical miles. The second strip chart covers New Georgia to New Ireland. The distance between Torokina, Bougainville and Rabaul, New Britain is a little over 200 nautical miles.

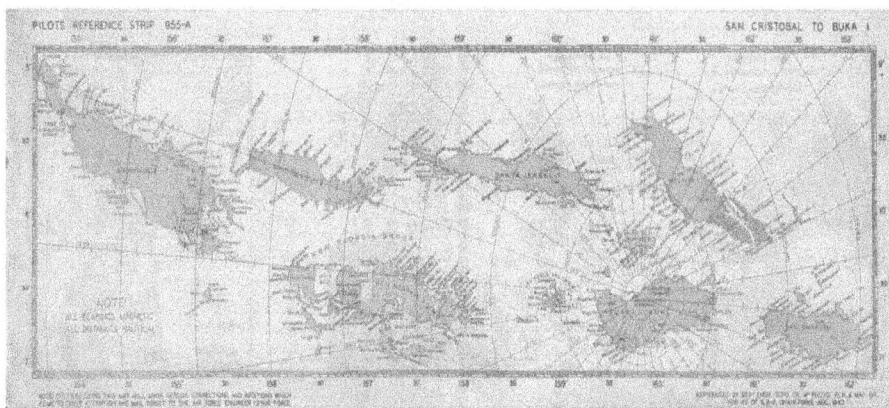

Pilots Reference Strip Chart – San Cristobal to Buka I. U.S. Government Chart.

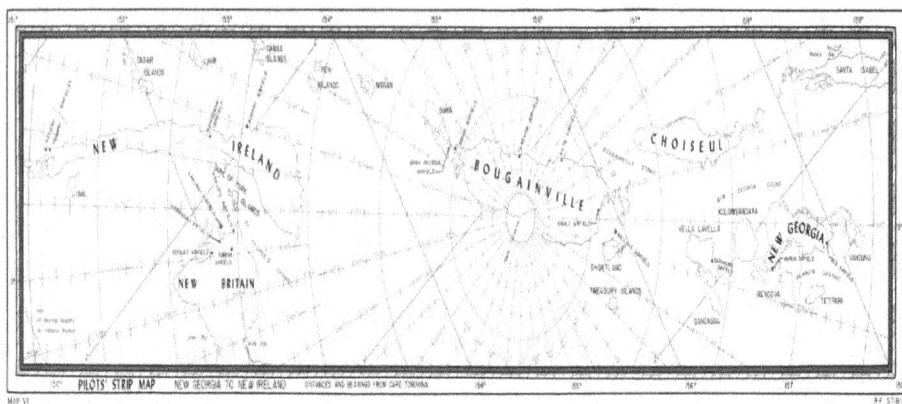

Pilots Reference Strip Chart – New Georgia to New Ireland. U.S. Government Chart.

As airfields were opened to the north, CASU-11 became a major repair shop, source of trained Sailors and parts supplier to the advanced aviation units. Often within minutes a pilot would fly in with a barely flyable aircraft, land, hop out, be pointed to a repaired aircraft and takeoff with an operational plane. On other occasions, a group of pilots would arrive via a large transport plane and then fly away as a group in repaired or recently received new aircraft.

SBD Parking Henderson Field. Navy Photo.

Appendix G reveals the process of getting a pilot ready to fly a mission.

December 1943

CINCPAC *Operations in Pacific Ocean Areas* report for December 1943: During the month of December, the principal missions of land-based aircraft in the South Pacific Area were: to maintain air supremacy over Bougainville Island and to the southeast; to contribute to the isolation and destruction of enemy forces on Bougainville by destroying his supplies and shipping; and to complete preparation for and commence attacks on the enemy's stronghold at Rabaul.

From 19 December on, both fighter sweeps and escorted bomber attacks were made on Rabaul. Our fighter sweeps usually were made with 40 to 50 VF [Fighters] at high altitude. The normal bombing mission included 18 B-24's and approximately 50 escorting fighters. The result claimed for these missions aside from damage caused by bombs, was 128 Japanese fighters destroyed to a loss of 20 of our planes. An estimate of enemy air strength in this area at the beginning and end of the month showed a decrease from 310 planes to 211 planes.

Carrier operations during 1943 started with only two combat carriers in the Pacific. By the end of the year the carrier strength had been augmented to 7 CV, 7 CVL, and 19 CVE, so that it was capable of, and was taking, decisive offensive action against enemy air and shipping centers beyond the range to which land-based planes could be brought.

The CV aircraft carrier was 875 feet long, carried 90 aircraft, and could travel at a maximum of 33 knots. A CVL was an aircraft carrier light. It was 635 feet long with 33 aircraft and could speed 31 knots through the water. And finally, the CVE was an aircraft carrier escort. The CVE was typically built upon a commercial ship hull, hence they were cheaper and built faster than the other two carrier types. The CVE was 495 feet long, carried 24 aircraft and had a maximum speed of 24 knots. The CVE was most often used to provide protection for convoys and slow- moving vessels. Admiral Nimitz welcomed every one of these ships that he could obtain for his fleet.

Wed. 1ˢᵗ Quiet. More rain. Had 12 to 4. My lip is getting better, thank goodness. Thurs. 2ⁿᵈ Quiet. Pay day yesterday. Have $176.00 on the books. Fri. 3ʳᵈ. Quiet. Sat. 4ᵗʰ Quiet. Sun. 5ᵗʰ Quiet. No Mass. Out on field. Mon. 6ᵗʰ Quiet. I haven't mentioned that it has rained for the last 8 nights. Not hard, but enough to make the show uncomfortable. Tues. 7ᵗʰ Quiet. Rained all day and all night. Miserable on the field. Just one big mud hole. Wed. 8ᵗʰ More rain. Quiet. Our tent is leaking like a sieve. 10 months is a long time for a tent. It's beginning to rot. Our bunks are all getting wet. I have a cough, and my lip is still sore. Also can't get rid of the fungus on my finger. Jock itch is much better, thank goodness. Boy, it's raining cats and dogs now. I guess the bad season is really here. Sure wish I was home, sitting before a nice fire with Olive. From the diary of Harry Hays, December 1943.

The rain was miserable. From November through December as much as 30 inches of rain would fall. Marine Corpsmen called all skin problems; such as the infection on Hays' finger, "tropical ulcers" and they could become serious. In some cases, the ulcer would damage the skin and tissue to the bone.

Thurs. 9ᵗʰ Quiet. More rain. We put another tent over the old one. Better now. I have a cold darn it. Fri. 10ᵗʰ Quiet. It has stopped raining and the sun is out again. My lip is still sore. Sat 11ᵗʰ Quiet. Boy, this is sure getting monotonous. Sun. 12ᵗʰ Quiet. Went to Mass. Mon. 13ᵗʰ Quiet. Tues. 14ᵗʰ Quiet. Wed. 15ᵗʰ Quiet. Thurs. 16ᵗʰ Quiet. Fri. 17ᵗʰ Quiet. Sat. 18ᵗʰ Quiet. Nuts!! Sun. 19ᵗʰ Quiet. Went to Mass. Lip and jock itch and finger all well. Mon. 20ᵗʰ Quiet. It is now 4:00 P. M. and the Condition Red has just sounded. First one in a long time. 4:25 Condition Green. Tues. 21ˢᵗ Quiet. Wed. 22ⁿᵈ Quiet. We have to put our mosquito nets up again. Too much malaria. Nuts!! Thurs. 23ʳᵈ Quiet. Went to confession. Fri. 24ᵗʰ Quiet. Christmas Eve. We were able to get 3 bottles of beer each today. 10 cents each. It's raining cats and dogs again. Also are allowed to have lights on until 2200 tonight. Some Christmas present! Went to Midnight Mass. Very nice- went to Communion. From the diary of Harry Hays, December 1943.

From 14 to 20 December, over 60 SBD sorties were flown each day against the Japanese on Bougainville. Each day the damage to airplanes by gunfire was slight to moderate. However, between launches rearming SBD guns and bomb bays, refueling the airplanes and normal flight maintenance, everyone kept busy.

Damaged SBD Dauntless. Navy Photo.

Recreational alcohol was officially prohibited on American warships throughout the war, but Sailors assigned in remote areas of the Pacific were sometimes given a small ration of weak beer with a 3.2% alcohol content, and later they could purchase limited amounts. The Marines provided enlisted men in more remote posts a ration of as much as two cans of beer per day.

This was intended to manage the amount of beer an individual could obtain at one time. Hays has mentioned in his diary that he had opportunities to buy a beer card, which had a limit of five beers, to use at the canteen. On special occasions leadership would open beer distribution for all Sailors, two cans per person at 10 cents apiece.

However, the higher your rank the more accessible alcohol was. Officers always had readily available booze. It was permitted for officers to organize a "wine mess" where all officers contributed equal amounts of money to purchase alcohol. When a new officer arrived, the value of the booze in stock was determined and divided by the number of members. This valuation set the new member's cost. The officers of CASU-11 always had a wine mess in operation.

149

During World War II, the U.S. Department of Agriculture ordered 5 percent of all beer production in the United States reserved for the troops. One interesting footnote: Many cans of American beer were painted olive drab to camouflage them from enemy aircraft.

It was generally admitted by all personnel on Guadalcanal that drinking was a way to alleviate boredom, raise morale and "overcome' fear."[87]

Admiral Nimitz rewarded a group of Marine aviators flying out of Guadalcanal with two cases of Old Crow whiskey. "Operations officer stuck the cases under his bunk bed. He would issue one fifth a night to the fighter pilots on Guadalcanal. There would be a one ounce jigger to give every pilot one drink. A couple of the men got two. We had a Guadalcanal cocktail, made of ice, unsweetened grapefruit juice and Old Crow. It was set up around a table. You'd go strung like a bow string, tired, nervous. As people drank, talk began and started a feeling of good fellowship. We'd ease up and by 8:30 or 9 everybody was asleep."[88]

The canteens of those at the front lines contained brandy, whisky or gin almost as often as water. Even Ernie Pyle finally took to drink, as he explained in a letter to his wife, going drunkenly to sleep "you wouldn't hear the guns and planes."[89] "In my outfit," James Jones said in his book, *The Thin Red Line*, "we got blind asshole drunk every chance we got" on what the troops termed *swipe*, an ad hoc distillate of sugar, canned fruit, potato peelings, and other such ingredients."[90]

In addition to legal alcohol, men built stills and made alcohol. Others used natural stills like coconuts. Medical personnel and Allied service members were happy to trade medicinal alcohol for war souvenirs. Astor reported that, "the tedium, broken only by routine missions, induced many aviators to seek release though alcohol. I watched Delaney, another gunner, who could find liquor anywhere, make a still [in the] back of the hut … It made about a gallon a day, just enough to take care of Delaney and his buddy. Another trick was to put a hole in a coconut and mix certain sugar and stuff with the coconut milk and when fermented you had a pretty strong drink. I tried 180 proof mixed with grape juice once, and once was enough. How those guys could drink that was beyond me."[91]

During his interview O'Flynn reported, "Well, they had bootleggers over there." He shared that his first drink of alcohol in his life occurred on Guadalcanal. Alcohol was always useful as trading material. O'Flynn continued, "One time we had a couple of Army guys in a jeep ... ran in a pothole with it and bent one of the springs on the main carriage of the jeep and it set and sputtered. They came by, and our Engineer Officer was there. They said, I need to get this fixed before I take the jeep back and turn it in. What would it take to fix it? We would have to take that spring off and heat it up and straighten it out because about six inches was bent straight up off the run of the spring. He says, could you do that? Well, here's the Engineering Officer, ask him. Well, he says I have a quart of booze here. If you will do it, I'll give you this quart of booze. I turned to the Engineer, Mr. Decker, said, what do you think of that? That sounds good to me. So we took the spring off, heated it up, straightened it out and put it back on. He gave us the booze and off he went and we had a big party."

After seven months on Guadalcanal, most of the men of CASU-11 had found recreational outlets of one kind or another to deal with the heat, boredom and stress. Work filled many hours, but there were still empty hours. For some, they drank a lot and hoped for the war to be over soon.

Some of the extra time was spent on camp and tent housecleaning. There were always coconut fronds and coconuts to remove from around the tents. During a dash to your foxhole these could be trip hazard. Sweeping the tent floor was a daily task as dirt and dust accumulated on everything. And finally there was always a button to be sown on a shirt or some laundry to do.

Cards and board games were popular. Books and old magazines were read and passed around. Every unit that had a couple hundred men had an outdoor movie theater. There was no way to provide a roof, so the theater typically had a covered screen and coconut logs for seating in the open air. Sound was difficult as it was always noisy from the wind blowing through the coconut fronds and the birds chirping. The sound of pounding rain on the covered screen also contributed to the noise. The Sailors did not care; they sat in the rain and watched the movie.

Camp Theater. From John McAteer.

The men also gambled to fill the time. If you did not have to send money to your wife or parents, then you had little use for your pay day funds. There was very little to buy. And, at least, gambling would burn of some of that unfilled time.

CASU-11 metalsmiths (Left to right, Leva, Lochyk, Parker and Nelson) enjoying cards. From David Wilson.

The men liked music. Vinyl records were available and played on manual, wind-up phonographs. O'Flynn shared, "We listened to old Tokyo Rose

152

every night. You know a bunch of Sailors, if there is something that can be had, one way or another, they would get it. From an aircraft, they built their own little receivers that would pick up Tokyo Rose." O'Flynn further remembered, "Old Tokyo Rose was our only entertainment out there. If you had a radio, every night she would have a show, play music, give all the propaganda, and try to break the morale of the people."

Athletics was an important part of daily life and baseball was a favorite sport. After arriving in a new location and all the tents and work sites were established, a baseball field was next. O'Flynn remembered it this way, "Well, there was no hours. We had work to do. When we got the work done, we played ball or something." Badminton and horseshoe pitching were also popular.

Sat. 25th Quiet Merry Christmas! Sun. 26th Quiet. Went to Mass. Mon. 27th Quiet. Tues. 28th Quiet. Wed. 29th Alert, but no raid. Thurs. 30th Quiet. Fri. 31st Quiet. Tonight is New Year's Eve. I hope I never have to spend another year's end like this one has been. All I can think of is home. It's terrible out here now. Every day it gets worse. No place to go nothing to do. A guy will go crazy with too much of this. From the diary of Harry Hays, December 1943.

January 1944

CINCPAC *Operations in Pacific Ocean Areas* report for January 1944: In the South Pacific shore-based air forces continued active and made consistent escorted bomber attacks on Rabaul. During January, there were 28 major day attacks and 5 major night attacks on land targets in the Rabaul Area. The completed attacks on this objective totaled over 2,000 sorties, or 56% of all sorties flown by South Pacific planes.

These missions were carried out in the face of strong enemy opposition and were handicapped by adverse weather conditions. Bad weather caused 12 major daylight attacks early in the month to be cancelled or diverted from Rabaul to other targets, and also resulted in the loss of a number of our planes. It was not until 22 January that sustained daily attacks could be carried out.

The result of the attacks during the month in destruction of enemy aircraft was on a scale never before approached in the South Pacific Theater. The total estimated by COMSOPAC to have been destroyed (503 aircraft) was nearly 175 more than in any prior month. Despite these losses there was no apparent reduction during the month in the number of enemy aircraft based at Rabaul. And while damage to the enemy installations occurred as a result of the repeated bombing attacks, all Rabaul airfields were operational at the end of the month. The intensity of anti-aircraft fire and the number of planes intercepting our attacks, however, appeared to be slightly less.

It is estimated by COMSOPAC that the number of aircraft at Rabaul was 211 on 1 January and 266 on 29 January. Obviously replacement aircraft were brought into Rabaul, the first 'Tojo,' an Army fighter, was encountered over Rabaul during the month.

As in the previous month, enemy attacks on the Solomon Island Bases in January were confined to night raids. It is estimated that the Japanese made 65 such sorties during the month losing 8 planes, 4 of them to our night fighters. These attacks, particularly during the middle of the month, were more destructive than usual. Our losses were 9 killed, 59 wounded, 6 planes destroyed and 10 more damaged.

Sat. 1ˢᵗ day of 1944. Quiet. It has rained continually for about a week. Sun. 2ⁿᵈ Quiet. No Mass. At field. Mon. 3ʳᵈ Quiet. Still Rainy. Tues. 4ᵗʰ Quiet. Another draft of 10 men to be drawn out of hat tomorrow. Bougainville. Boy, oh Boy!! Wed. 5ᵗʰ Quiet. Well, I was lucky again- thank goodness! Griswald got caught, and after he and Bill built such a nice tent- too bad! Well, Griswald doesn't have to go. Malaria twice, and sore on leg looks too bad on his record. Thurs. 6ᵗʰ Quiet. Fri. 7ᵗʰ Quiet. Sent $222.00 to my sweet wife. Sat. 8ᵗʰ Quiet. Rain. Sun. 9ᵗʰ Quiet. Didn't go to Mass because raining cats and dogs. Mon. 10ᵗʰ Quiet. More rain. Won $9.75 at poker. Tues. 11ᵗʰ Quiet. Rain. Wed. 12ᵗʰ Quiet. Rain. Boy, what a mud-hole! Thurs. 13ᵗʰ Quiet. More rain. Fri. 14ᵗʰ Quiet. Had regular hurricane last night. Wind blew top of hut off out at strip. Also blew hood off our tent. My bunk is soaked. What a place! Sat. 15ᵗʰ Quiet. More rain and wind. Sun. 16ᵗʰ Quiet. No Church. Had to work. That's 3 times I've missed in a row. Rained practically all day. Mon. 17ᵗʰ Quiet. Had rained off and on all day. It's now 4:00 P.M. And it is raining *as hard as I have ever seen it. By the tub full! You can hardly see 10 feet away from you. Tues. 18th Quiet. More rain! Wed. 19th Quiet. No rain today. Sun is out, and it's just blazing - you can see the steam rising from the ground. Thurs. 20th Quiet. Hot- and how! Fri. 21st Quiet. Sat. 22nd Quiet. They're sending Ginther out. Too much malaria. One time right after another. Sun. 23rd Quiet. Went to Mass. Mon. 24th Quiet. Tues. 25th Quiet. Mosquitos are bad again. Wed. 26th Quiet. Thurs. 27th Quiet. Fri. 28th Quiet. Sat. 29th Quiet. Sun. 30th Quiet. No Mass. Had to work. Mon. 31st Quiet. Steele, one of the mechs, went berserk today.* From the diary of Harry Hays, January 1944.

An average January would have had 17 inches of rain at Henderson Field. CASU-11 reported the daily average number of takeoffs and landings for January 1944 as follows: 13 fighters, 89 bombers, and 77 transports. The CASU crew had 35 aircraft under their charge for maintenance.

Steele would have been taken to MOB-8, or Mobile Hospital No. 8, which was constructed on Guadalcanal in April 1943. In August 1943 it was commissioned with 1,290 beds. During 1943, 2,208 patients were admitted for some form of psychoneurosis. By December 1944, this hospital had treated 39,395 patients. The hospital was a couple hundred yards from CASU-11's campsite.

February 1944

CINCPAC *Operations in Pacific Ocean Areas* report for February 1944: The major efforts of the first part of the month were devoted to increasingly heavy attacks on the enemy's aircraft at Rabaul, and to preparations for the amphibious landing on the Green Islands. On 15 February, the Green Islands landing was successfully accomplished, and by the 19th, the Japanese air forces at Rabaul had been largely destroyed. Overall Japanese activities in the South Pacific were largely confined to attempts to counter the aggressive moves of our forces. These defensive efforts were notably unsuccessful.

The number of Allied planes based in the Solomon Islands during the month remained generally constant. At the end of the month approximately 740 aircraft were on hand with 230 being SBD's and

TBF's. As the capacity of our forward bases to accommodate planes increased, additional short ranged planes were based at each. Thus, by the end of the month, approximately 60% of our fighters, 50% of our SBD's, and 72% of our TBF's were based within 215 miles of Rabaul.

During the period from 19 December (when attacks on the Rabaul Area commenced) to the end of January, the assault by land-based aircraft proceeded with growing intensity. In this time it is estimated that our forces had destroyed 656 Rabaul based enemy planes. As a result of this constant pressure, enemy fighter interception had somewhat decreased, but still all of his [Japanese] airfields remained serviceable.

During the first 19 days of February, the tempo of our attack continued to increase. In this time over 3,000 sorties were flown against Rabaul. While this severe bombing campaign was in progress our fighters were continuing to take a heavy toll of enemy fighters. Particularly successful days were 9 February, when 22 enemy planes were destroyed, and 10 February, when 35 were shot down. By 20 February our planes had accounted for 207 enemy planes, an average of over 10 per day since the first of the month. The final blow came on 19 February when 27 enemy interceptors were destroyed while only one of our planes was lost.

After the middle of February the entire situation at Rabaul changed. Interception of our planes over Rabaul by enemy aircraft ceased after 19 February. For the remainder of February maximum bomb loads were carried and dropped not only on airfields and installations, but also on enemy supply and barrack areas at Rabaul Town. Thirty major strikes were made during the last 11 days of the month. On a number of these attacks fighters as well as bombers were used to carry bombs. In all nearly 2,900 tons of bombs were dropped on targets in the Rabaul Area during the month. Of 28 bombers destroyed by the enemy during February attacks on Rabaul Area targets, 15, or approximately 53%, are believed to have been lost to anti-aircraft fire. For each plane destroyed, an additional 15 were damaged.

Tues. 1ˢᵗ Quiet. Wed. 2ⁿᵈ Quiet. Booten is at MOB-8 for a nervous breakdown. It has rained for the last 3 days, and is starting again right now. Thurs. 3ʳᵈ Quiet. Rain. Fri. 4ᵗʰ Quiet. Rain. Sat. 5ᵗʰ Quiet. Sun is out. Sun. 6ᵗʰ

Quiet. Went to Communion. Hot again. Mon. 7th Quiet. Had 8 to 12. Poured rain. Tues. 8th. Quiet. Still pouring rain. Wed. 9th Quiet. More rain. Thurs. 10th Quiet. More rain. Fri. 11th Quiet. More rain. Sat. 12th Quiet. Sun is out. Sure feels good. Sun. 13th Quiet. Went to Mass. Hotter than heck. Mon. 14th. Quiet. Sick Bay is taking malaria tests of every man. Just heard the Skipper cracked up in an SBD. Took him to MOB 8. Also CASU #14 is relieving us at the Tower. Must be something brewing. From the diary of Harry Hays, February 1944.

Lt. Cmdr. Schlossbach, Commanding Officer of CASU-11 and Henderson Field, was the Skipper referenced. He was a submarine captain during World War I and a South Pole explorer. Over 50 years old, he volunteered for duty in World War II, became a qualified Navy aviator, and was assigned to Guadalcanal. However, because of a medical condition, his left eye was surgically removed. Consequently, he was only allowed to fly with another qualified pilot.

In spite of being restricted to multi-pilot aircraft and with almost tragic results, Schlossbach took an SBD for a joy ride. He lost control of the plane as it landed and it broke the wing off a parked aircraft. The wing then flipped through the air and jammed against Schlossbach's chest. All of his ribs were broken. Fortunately he was strapped in or he would have died. The skipper was taken to the nearby medical facility, MOB-8, where he remained in treatment until March 1944 when he was shipped home. Schlossbach fully recovered and was eventually stationed in Argentia, Newfoundland, for the remainder of the war.[92]

CASU-14 had been working on Munda Field, New Georgia, Solomon Islands. However, after 18 February, they split their manning for the remainder of their time overseas between Munda Field and Guadalcanal. Their arrival on Guadalcanal must have started the rumor mill with talk about when CASU-11 might go home or if CASU-11 would be reassigned to another South Pacific location.

Tues. 15th Quiet. Wed. 16th Rained all night. A pipeline from the mess hall broke and flooded our fox hole. We had to fill it in today on account of the stink and mosquitos. Thurs. 17th. Quiet. Fri. 18th Quiet. CASU #14 Took over today. Sat. 19th Quiet. Well, I'm back in the shop again, darn it! Sun. 20th

Quiet. Went to Mass. Have been put on the line crew. I hope it's better. Mon. 21ˢᵗ Quiet. Worked on tool box today. Go on line tomorrow. Tues. 22ⁿᵈ Quiet. Wed. 23ʳᵈ. Quiet. Thurs. 24ᵗʰ Quiet. Fri. 25ᵗʰ Quiet. Sat. 26ᵗʰ Quiet. Sun. 27ᵗʰ Quiet. Went to Mass. Mon. 28ᵗʰ Quiet. Got called on duty at the field operations office. In charge of transportation. It's a pain in the neck! Tues. 29ᵗʰ Quiet. Boy, what a job! From the diary of Harry Hays, February 1944.

February was another wet month. Historically the maximum rainfall was 17 inches. In February 1944 the actual rainfall was 16 inches. This weather was problematic when the Allies were trying to keep continuous pressure on the Japanese at Rabaul. Maintenance and repairs continued but at a slower pace. Wet slippery tools and parts fell into the mud. The gusty winds reduced visibility. This all combined to make difficult repairs almost impossible.

A section of CASU-14 took over the responsibility of operating Henderson Field, and, as a result, all CASU-11 personnel returned to working only on aircraft. Hays' comment that he had, "been put on the line crew" meant he would not work at the metalsmith shop but rather out on the flight line near the runway. He and his men performed between flight checks; such as, tire pressure and oil checks, refueling and reloading guns and bomb racks. The flight line would be intermittently wet and muddy or windy and dusty. The steel Marsden matting under foot was hard on the knees and legs. It was lousy working conditions.

This change in the work assignments for CASU-11 and 14 over the last six weeks was driven by the doubling of fighter and bomber aircraft on Guadalcanal and the Russell islands.

Starting in January and ending in March there was maximum effort by land-based aircraft against Rabaul. In January some "maintenance" no-fly days were taken because the maintenance requirements involved more man hours than were available. On 17 February ten SBDs and five other aircraft were damaged by enemy gunfire and returned safely. On 29 February all planes returned with two SBDs each having several machine gun bullet holes.

The next picture shows the type of damage that the metalsmiths repaired.

Damaged rudder. Navy Photo.

March 1944

CINCPAC *Operations in Pacific Ocean Areas* report for March 1944: The month of March was a period both of consolidated advances made in February, and of preparation for future operations. Heavy air assaults on Rabaul continued with a goal of neutralizing the remaining enemy bases in the area of the Northern Solomon Islands and the Bismarck Archipelago. In fact, air attacks on the principal enemy base, Rabaul, were heavier than in any previous month.

During March the number of aircraft in the Solomons was increased, principally in the number of heavy bombers available. By the end of the month, 14 fighters, 20 SBD dive bombers, and 20 TBF torpedo planes had been moved north to the new field in the Green Islands.

By the first of March, enemy aerial opposition to our attacks on Rabaul had virtually stopped. Contact with Japanese fighter planes was made only rarely during the month, however, anti-aircraft fire at Rabaul remained intense and all its airfields remained usable in spite of our heavy bombing. The extent of the March effort (3,262 sorties) surpassed that of any preceding month. The daily average of bombs dropped on Rabaul targets exceeded 100 tons. By the end of March, much of the town of Rabaul was burned out, and harbor installations were badly battered.

To prevent further use of the airfields by the enemy, further bombing was necessary. Over 700 tons of bombs were expended on these in March, and by the end of the month, they were all reported to be unserviceable. Incident to these operations six enemy aircraft were destroyed in the air, and 10 on the ground. These figures contrast sharply with those for February when the Japanese were actively defending Rabaul in the air. After this point it became possible to use fighter aircraft as bombers. They usually carried 500 pound bombs, and with a little practice they were able to achieve bombing accuracy comparable to that of dive bombers.

Wed. 1ˢᵗ Quiet. Thurs. 2ⁿᵈ Quiet. Pay day. Have $99.00 on books. Fri. 3ʳᵈ Quiet. Sat. 4ᵗʰ Quiet. Raining again. Sun. 5ᵗʰ Quiet. Couldn't go to Mass on account of work. Mon. 6ᵗʰ Quiet. More rain. Tues. 7ᵗʰ Quiet. More rain. Wed. 8ᵗʰ Quiet. More rain. Thurs. 9ᵗʰ Quiet. More rain. Fri. 10ᵗʰ Quiet. Sun is out. Sat. 11ᵗʰ Quiet. Sun. 12ᵗʰ Quiet. No Mass as I had to work. Mon. 13ᵗʰ Quiet. First day off in 2 weeks. It is now 4:00 P.M and has started to rain cats and dogs. The mess hall has water all over the deck 6" deep. Tues. 14ᵗʰ Quiet. Chief Moore made Warrant last Sept. Just found out about it. Is going back home. Wed 15ᵗʰ Quiet. I have working party today. Thurs. 16ᵗʰ Quiet. Fri. 17ᵗʰ Quiet. Sat. 18ᵗʰ Quiet. Sun 19ᵗʰ Quiet. Went to Mass. Mon. 20ᵗʰ Quiet. Tues. 21ˢᵗ Quiet. Wed. 22ⁿᵈ Quiet. Thurs. 23ʳᵈ Quiet. Fri. 24ᵗʰ Quiet. Sat. 25ᵗʰ Quiet. The wind sure blew last night. Almost a hurricane! We darn near lost our tent. Sun. 26ᵗʰ Quiet. Went to Communion. Mon. 27ᵗʰ Quiet. Tues. 28ᵗʰ Quiet. Wed. 29ᵗʰ Quiet. Thurs. 30ᵗʰ Quiet. Got drunk helping Gris celebrate making Chief. Fri. 31ˢᵗ Quiet. From the diary of Harry Hays, March 1944.

Seabees working near CASU-11 reported the March 1944 rainfall amount was 12.5 inches. All of these "quiet" non-raid days happened because the air assault on the Japanese airfields between Guadalcanal and Rabaul, and upon Rabaul itself, were taking a serious toll on Japanese aircraft, in the air and on the ground. The work for the men of CASU-11 was unabated. In March

1944 there was a peak of 2,490 Dauntless SBD bomber sorties flown over Rabaul.

April 1944

CINCPAC *Operations in Pacific Ocean Areas* report for April 1944: Medium, light, and fighter bombers flew 4,027 sorties against Rabaul, a number greater than in any previous month. By the end of April, it was reported that Rabaul Town had only one building standing for every five acres of area. All known supply areas surrounding Rabaul had been heavily attacked for over two months and by the end of April appeared to have been largely ruined. Particularly concentrated attacks were made on all enemy trucks and vehicles that could be located. As a result, a considerable number were destroyed.

While these large scale air attacks were successful against buildings and supplies above ground, the results against the five airfields and against anti-aircraft gun positions were not as effective. In spite of 1,000 tons of bombs dropped on airfields, the enemy doggedly continued to repair them, and at the end of April they appeared at least as serviceable as they were when the month started. Although a reduction in enemy anti-aircraft fire was reported, the fact that 17 of our planes were lost during April in the Rabaul Area, due almost entirely too anti-aircraft fire, indicates that this defense was still strong.

Sat. 1ˢᵗ Quiet. Sun. 2ⁿᵈ Quiet. Went to Mass. Mon. 3ʳᵈ Quiet. Tues. 4ᵗʰ Quiet. Tomorrow is pay day. Have $146.00 on books. My pay never seems to come out right! Wed. 5ᵗʰ Quiet. Thurs. 6ᵗʰ Quiet. Fri. 7ᵗʰ Quiet. More rain. Sat. 8ᵗʰ Quiet. Rain. Sun. 9ᵗʰ Quiet. Went to Mass. Mon. 10ᵗʰ Quiet. Tues. 11ᵗʰ Quiet. Wed 12ᵗʰ Quiet. Took competitive test. Thurs. 13ᵗʰ Quiet. Fri. 14ᵗʰ Quiet. Passed exam. Sat. 15ᵗʰ Quiet. More rain. Sun. 16ᵗʰ Quiet. Mon. 17ᵗʰ Quiet. Took other test. Tues. 18ᵗʰ Quiet. Wed. 19ᵗʰ Quiet. Thurs. 20ᵗʰ Quiet. Fri. 21ˢᵗ Quiet. I passed test. Now have to wait until it's signed. Sat. 22ⁿᵈ Quiet. Sun. 23ʳᵈ Quiet. Went to Mass. Mon. 24ᵗʰ Quiet. Tues. 25ᵗʰ Quiet. Wed 26ᵗʰ Quiet. Thurs. 27ᵗʰ Quiet. I feel terrible – I made Chief as of today and tomorrow or next day am being transferred to C.A.S.U. #9 at Russell Islands. It wouldn't be so bad if someone were going with me, but I have to go all alone. I sure do miss my wife tonight. Fri. 28ᵗʰ Quiet. Almost got out of it, but not quite. Damn Snyder anyway! Rate and transfer go into effect the 1ˢᵗ *of May. Sat. 29th Quiet. Nothing new. Sun. 30th Quiet.* From the diary of Harry Hays, April 1944.

The Russell Islands lie approximately 50 miles west northwest of Guadalcanal. It was not surprising that Hays was reluctant to go to the Russell Islands without anyone else from CASU-11. Life in CASU-11 built relationships that lasted a lifetime. When men are thousands of miles away from family and friends and experience harsh and dangerous conditions, they bond tightly. "Band of brothers" does not come close to describing the feelings most men had for each other.

In 2015 John Parker, son of Aviation Metalsmith Second Class Frank Parker, wrote a blog about his father. He remembered his Dad telling him, "when CASU-11 got home from Guadalcanal, the guys all went their separate ways. My Dad stayed in California after meeting and marrying the love of his life, my mom. More than forty years later, two weeks after retiring, Dad got a call from the DMV [Department of Motor Vehicles]. They confirmed his name and asked him to stay near his phone. A few minutes later, one of his old buddies came on the line and asked, 'Is this Frank?' When my dad said 'yes, 'the man started crying. For some time he had been searching to find all the members of CASU-11's metalsmith crew and had put together a reunion at the Lake of the Ozarks. My dad was the last member to be found and he and my mom caught a plane to meet the group.

"Obviously, it was a great time for all of them. The bond was so strong; they continued to meet once and sometimes twice a year for the rest of their lives. They called each other weekly and made private visits to each other's homes. When they slowed down, their children provided transportation and enjoyed the group's visits and stories. They were humble men, but true patriots and heroes."[93]

On 17 April 1944, Lt. Cmdr. Carleton Pike became the second Commanding Officer of CASU-11. He was a naval aviator and had already served 19 months in the South Pacific. He served for 3 months in this new position and then he went home. His time in the war zone was over. More details about Pike are provided in Appendix C.

During April 1944 the Seabees reported 9.5 inches of rain.

May 1944

CINCPAC *Operations in Pacific Ocean Areas* report for May 1944: Operations of land-based aircraft in the South Pacific Area were much the same as in April. The principal functions were to tighten the blockade around the remaining enemy garrisons and to destroy enemy personnel and equipment wherever found. To attain these objectives the number of combat sorties (6,196) flown over the Solomons increased over the April total.

The principal development in the Rabaul campaign during May was the establishment of a continuous patrol over the area night and day by fighters, B-25s [Army long range bombers], or Ventura night fighters. This patrol was useful not only for reconnaissance, but helped to deny the area to enemy aircraft. Aside from this development, the inexorable bombing of gun positions, airfields, and all enemy transportation facilities continued. Total sorties and bomb tonnages decreased by approximately one third as compared to the April figures.

Enemy anti-aircraft fire at Rabaul remained effective during May, being responsible for the loss of 15 of our aircraft. SBDs continued to bomb these positions while TBFs usually hit other targets. While enemy offensive action against our South Pacific bases has steadily decreased, May was the first month since the establishment of these bases that no attacks were made on them by enemy aircraft.

Hays had now joined CASU-9 located just to the north of Guadalcanal in the Russell Islands. The remainder of his diary is provided in Appendix B.

Early in May, O'Flynn injured his arm. "I got my arm burnt and ligament pulled in my elbow…it was the only time I was ever hurt," he said. "I went to sick bay. They had a tent there, sick bay in it, the old doctor says, 'Sit down here in this chair and put your arm up on the table.'" O'Flynn was instructed to straighten his arm but couldn't. The doctor attempted to straighten his arm as well but met the wrath of an injured and hurting man. "I was sitting there with tears a rolling down my cheeks. I jumped up and he jumped back against the wall of the tent. He said, 'I think I better send you to the real hospital,' because I was going to get him off that arm. He was killing me! Good thing I did not hit him because I was going to."

About three weeks later, over in the hospital, O'Flynn's arm had healed to the point where he could use it some. "I was out one day there sweeping the grinder where they train. The old doctor came by, and he said, 'Hey O'Flynn, come over here. What the hell are you doing out here?' I said, 'It's so boring, I came out here. [I've] been telling you I am ready to go back to my unit.' He said, 'I know a little bit more than you do about when you are ready to go back to your unit. You throw that broom down, you get in that barracks and behave yourself.' So that was the end of that." O'Flynn was released to full duty on 2 June 44.

Charles Lindbergh, a controversial isolationist and anti-war personality, was one of the people the members of CASU-11 often spoke about upon returning home. While Lindbergh did not interact directly with CASU-11, everyone knew he was in the area. It is a certainty CASU-11 personnel may have observed his takeoffs and landings and may have seen him on the island.

Although his politics did not seem to fit his actions, Lindbergh was very involved in improving the Corsair fighter, considered by the Navy and Marines as their finest aircraft. Even though he was a civilian, he was qualified to fly the F4U, a version of the Corsair.

He went through several weeks of combat training with the Marines. On 22 May 1944, while Lindbergh was at Espiritu Santo, he flew his first combat mission with VMF-222, a Marine fighter squadron.

By June Lindbergh was operating out of Guadalcanal, teaching and demonstrating his theories on how to improve Corsair performance and flying missions with the Marines and Navy. It is estimated that Lindbergh flew over fifty missions with the Marines, Navy, and later with the Army.

Lindbergh was convinced that the powerful Corsair could carry much more ordnance than it was normally fitted with. On September 8, he flew an aircraft with a specially modified belly rack that carried a 2,000-pound bomb. Taking off from Roi Island, he flew to Wotje Island, and dropped the weapon on a Japanese installation.

"On 13 September 1944, Lindbergh had his plane, an F4U-1D fighter bomber, loaded with one 2,000 pound and two 1,000-pound bombs. Using up almost every foot of runway, he managed to pull the shuddering plane into the air and proceed to Wotje where he dropped the heaviest load carried by a single fighter up to that time."[94]

June 1944

CINCPAC *Operations in Pacific Ocean Areas* report June 1944: South Pacific Areas saw little activity except for routine neutralization of enemy positions and patrols that continued as in the past. Tactical aircraft for the defense and logistic support of the South Pacific Area were now to be stationed at Guadalcanal, Espiritu Santo, New Caledonia, New Zealand, Fiji, the Ellice and Samoan Islands. The airfields in the Russell Islands, although not in active use, were to remain available for staging carrier air groups should this become necessary. Many fields were to be maintained only for emergency use, and others were to be abandoned. On 15 June, 608 Allied tactical planes composed the airforce of the South Pacific Command.

The principal feature of the month was the continuing maintenance of a 24-hour combat air patrol over Rabaul rather than the use of constant bombing as a means of preventing the enemy's use of the airfields there.

On 3 June 1944, O'Flynn came down with his third case of malaria. It was his most serious bout with this fever, and the doctors ordered him to the hospital. O'Flynn said, "New Caledonia is where there was a big hospital there in the Pacific, but on the fringe of the war zone. The doctor says we will have to wait a couple of days to get a ship going to New Caledonia, and we will send you up there to the hospital." Prior to leaving Guadalcanal, O'Flynn met some Marines from the First Marine Division, where his brother was assigned. The Marines told him how to get mail to his brother. O'Flynn made contact, and planned to meet his brother in New Caledonia.

On Thursday 1 June, and unknown to O'Flynn's doctor, the USS Tryon (APH-1) had arrived on Guadalcanal with its next port of call at New

Caledonia. O'Flynn was transported from the hospital, boarded the Tryon and on the afternoon of Tuesday, 6 June checked in to the Mobile Hospital Number Five, New Caledonia. A few days later the Marines brought O'Flynn's brother to the hospital. "We visited a while, it was the only time I saw him [during the war] - he survived the war. I had nine brothers. Eight of us was in the service. My brother was also in Marines in San Diego while I was there, so he came over a couple of times, while we were in boot camp. I did not see him again until '44, it was June '44 when I saw him over in New Caledonia." O'Flynn remained in the hospital until 10 July. He would not know that the majority of the men of CASU-11 had already departed Guadalcanal for the U.S until his return to the CASU-11 campsite.

On 30 June 1944 most of CASU-11 departed Guadalcanal on the USS Rochambeau. The war had moved away from Guadalcanal. Wilson, Little and West were part of this group headed back to the continental United States. USS Rochambeau with CASU-11 men on board arrived in San Francisco on 20 July 1944, anchored in the bay at Berth 7 and then off loaded all passengers at Pier 31.

After arriving at the Receiving Station, San Francisco, California, Wilson was ordered to CASU-6 located at NAS Alameda, California. On 13 December 1945, he ended his service with the U.S. Navy.

Little was ordered to NAS Corpus Christi, Texas, where he continued his Navy service. In June 1962 he completed twenty years of honorable service. West was ordered to CASU-50 at Pasco, Washington. On 6 December 1946 he ended his service with the U.S. Navy.

As men began to depart CASU-11 their service records were reviewed and updated by the CASU-11 personnel office. Many noticed the following stamped entry in their record,

Entitled to wear Asiatic-Pacific Ribbon in lieu of medal.
AUTH: Gen'l Ord. 194 & 207

This was acknowledgement that the service member had earned the Asiatic-Pacific Campaign Medal. The Asiatic-Pacific Campaign Medal was

166

authorized on November 6, 1942 by Executive Order No. 9265 signed by President Franklin D. Roosevelt. The Medal was issued to commemorate the service performed by personnel of the Navy, Marine Corps, and Coast Guard who served during the period in the area designated and met the prescribed condition."[95]

CASU-11 service members were eligible for this ribbon/medal because they served in Operation Cartwheeel -- Consolidation of Solomon Islands, 8 February 1943 through 15 March 1945. They were also eligible to receive an engagement star. The determination of the correct number of engagement stars for CASU-11 members has been a problem. Appendix D provides more information.

Asiatic-Pacific Campaign Ribbon with one Engagement Star. Cover has color picture of this medal.

July 1944

CINCPAC *Operations in Pacific Ocean Areas* report July 1944: There were no offensive operations in the South Pacific, and there was some reduction in our facilities in the Solomons Area. The pattern for air operations in July continued much as in June. Patrols and strikes were carried out at Rabaul and the Japanese garrisons were harassed, bombed, and blockaded throughout the entire area. Our losses during the month in this area decreased to 11 planes as compared to June losses of 18.

The summer of 1944 was a period of transition for the aviation commands scattered across the Solomon Islands. Changing orders and transfers were actively ongoing for many personnel and commands. Lt. Cmdr. Reinhart Vogt's transfer typifies the somewhat chaotic scene. On 25 June 1944, Vogt had orders to leave CASU-12 on Bougainville and report to CASU-9. On 26 June his orders were changed, directing him to proceed to CASU-11 on Guadalcanal and become the new commanding officer. In a 26 June letter to his wife Vogt said, "Hadn't left yet for my new assignment so they caught me in time. Don't know how I will like the new location [Guadalcanal] as it is where I spent last Christmas and New Year's which wasn't a very pleasant spot to spend those holidays. Maybe I can fix things up and bear it until my tour of duty out here is over."

Vogt's 3 July 1944 letter home shared his optimism on taking his responsibilities at CASU-11, "Leaving in the morning for my new assignment so may be quite busy for a spell. I hope my new unit is half as good [as my old one] and I'm sure I can make it a happy ship. Will now have to make it the number one unit in these parts."

On 5 July 1944 Lt. Cmdr. Vogt relieved Cmdr. Pike and became the third Commanding Officer of CASU-11. His 6 July 1944 letter shared "On my new assignment and put the officer I relieved on a plane this morning for the states. He was quite anxious to get back as he had been out here nineteen months. Believe I will like my new assignment as I see some constructive work that can be done. That always makes one's work interesting and makes the time fly by more rapidly."

Then Vogt made commander and heard from senior officers that he was being considered for another job. During his first week on the job he watched CASU-11 depart on various ships. He knew there was another change already in the offing. Vogt stated, "Things are beginning to break for me and tomorrow or the next day I take over command of CASU-41. Haven't put my old CASU-11 on the boat yet but soon. I'll only have it [CASU-41] a short time undoubtedly when I will take command of another bit larger unit and combine the two a little later when I am to take command of all aviation activities on the island. At least that is what the staff has on their schedule so now I'm working on a lot of consolidation."

In his 5 August letter to his wife, "Well I got my orders today so I will be leaving here in a few days by plane." Given the above letters expressing what Vogt thought was going to happen, the following is what actually occurred:

On 22 June 1944 Vogt left CASU-12 at Bougainville.
On 5 July 1944 he assumed command of CASU-11 on Guadalcanal for 17 days.
On 22 July 1944 he assumed command of CASU-41 on Guadalcanal for 14 days.
Finally, on 5 August 1944 Vogt departed Guadalcanal with orders to COMAIRPAC Pearl Harbor, Hawaii, for staff duty.

Just prior to Cmdr. Vogt's departure from CASU-12, he received the Bronze Star. "The Bronze Star Medal was awarded to recognize minor acts of heroism in actual combat or single acts of merit or meritorious service either in sustained operational activities against an enemy or in direct support of such operations."[96]

Meanwhile, Lt. Cmdr. Phillip Allen Jr. arrived and assumed command of CASU-11 on 22 July 1944. He was the commanding officer for only 12 days since most of CASU-11 was departing Guadalcanal for home during August 1944. Allen then executed his next set of orders to proceed to Pityilu, Papua New Guinea, where he become the commanding officer of CASU-42 on 27 August 1944.

This rapid and somewhat chaotic movement of airplanes, ships and personnel was a sign of Allied success in the South Pacific. The war was moving northward, ever closer to Japan.

On 10 July O'Flynn returned to CASU-11 from the hospital in New Caledonia and learned that they were all going home. He also learned that the bulk of CASU-11 was already underway and most of the way home.

On 13 July USS Rochambeau was 1,100 miles southeast of Hawaii and 2,200 miles southwest of San Francisco in high waves and hurricane force winds. On this day the deck log of the USS Rochambeau at 1600 [4:00PM] reported that "the ship passed through the center of a small tropical hurricane, with winds reaching a maximum velocity of 68 knots, and a minimum barometer reading of 28:20 inches. During the storm the hydraulic steering gear failed and control was shifted to hand steering. The broken hydraulic steering gear was repaired by 1815 [6:15PM], and steering control was shifted to the bridge." For the men of CASU-11, this meant rough seas of 10 to 14 feet and a ship rolling heavily for at least two days, a very uncomfortable ride.[97]

O'Flynn was placed on the next ship leaving Guadalcanal. On 14 July 1944, O'Flynn arrived at the Receiving Station Noumea, New Caledonia and on 29 August 1944 he arrived at the Receiving Station San Francisco, California.

On 25 July 1944 McAteer departed Guadalcanal. After some deserved leave, he attended two lengthy schools. This prepared him for duty on the USS Thetis Bay (CVE-90) where he served from 29 May 1945 to 10 May 1946 as the Supply Officer. On 17 August 1946, he exited the U.S. Navy.

O'Flynn remembered their arrival at San Francisco and Port Hueneme very well, "We got back in August of '44 and anchored out in San Francisco Bay, and I guess they forgot us two or three days there. Finally, they showed up and picked us up again and took us to Port Hueneme, you see. Liberty denied - hadn't had liberty since 1942, no liberty at night."

CASU-11 personnel on Guadalcanal continued to depart during August 1944. Finally, late August 1944 the last men of CASU-11 left Guadalcanal. Parker was in this group and upon arrival at the Receiving Station, Treasure Island, San Francisco, California, in September 1944, he received orders to CASU-6 at NAS Alameda, California. Parker left the Navy in June 1945. While most of the remaining men of CASU-11 received orders to other Navy commands, some of them had completed their enlistments and they were given orders home.

Part Four - Port Hueneme and Thermal, California

September - October - November - December 1944 - January - February 1945

CINCPAC *Operations in Pacific Ocean Areas* report: There were no offensive activities in the South Pacific. CINCPAC reports ceased coverage of the Southwest Pacific December 1944. CINCPAC reports will resume March 1945.

On 10 September 1944, approximately 100 men were still assigned to CASU-11. They were transferred, presumably by train, from the Receiving Center in San Francisco to the ACORN Assembly and Training Detachment (AATD), Port Hueneme, California.

"An ACORN Assembly and Training Detachment was established to the northwest of the Port Hueneme harbor adjacent to Silver Strand Beach. It covered approximately 141 acres with a capacity for 3,080 personnel."[1] Similar to CASU-5 in San Diego, the ACORN Assembly and Training Detachment collected and trained CASU personnel.

While CASU-11 was at the ACORN Assembly and Training Detachment, the goal was to increase the manning level of CASU-11 to 800 men and 50 officers and to train them to repair and maintain the PB4Y-2 Privateer, a new long-range, reconnaissance bomber. This new aircraft was manufactured at the Convair/San Diego plant, 180 miles south of Port Hueneme. This proximity created a perfect training opportunity at the plant for CASU-11

Sailors. Many CASU-11 Sailors also attended the PB4Y-2 Line Maintenance School at Naval Air Station Hutchinson, Kansas.

O'Flynn also remembered a very important event that happened to him after arriving at Port Hueneme. "When they got us back to Port Hueneme to establish another unit, they took, I think it was about 13 of us out of the original bunch and probably about 60 or 70 more people." A few days later O'Flynn recalled being stopped by the executive officer of CASU-11. "He said, 'What are you doing here?" I said, 'I've been reassigned to CASU-11 again." He said, "They're leaving here in January to go back to back out in the Pacific; you don't want to go out there, do you?'

"I said, 'I'd like to get on that highway 101 and get on a highway that leads off east as far as I can go get away from the Pacific.' He said, 'Well, we will see what we can do about it.' Four days later I had a set of orders to Naval Air Station Jacksonville (Florida). The Executive Officer's name was Harmon, Lt. Harmon."

During September 44 Lt. R.B. Torian was the Acting Commanding Officer. He was probably the most senior officer assigned to CASU-11 at that time.

On 25 October 1944 CASU-11 along with all CASUs permanently based outside the continental United States or Hawaii were notified that they would be redesignated as **Combat** Aircraft Service Units (**F**). The F signified forward. Therefore CASU-11 became CASU(F)-11 or Combat Aircraft Service Unit Forward. Although CASU-11 was presently located within the continental United States, this was temporary. All CASUs permanently based inside the continental United States and Hawaii retained their command title as **Carrier** Aircraft Service Units (CASU).

In early December 1944, CASU(F)-11 left Port Hueneme for the Naval Auxiliary Air Station, 29 Palms located in 29 Palms, California. After a few days they proceeded to Naval Air Facility (NAF) Thermal, California, and arrived on 7 December 1944.

M.L. Shettle shared the following information:

"During the first year of U.S. involvement in World War II, the Army hurriedly established a training center in the California desert, 25 miles southwest of Palm Springs. General Patton trained his army here in preparation for Operation Torch - the invasion of North Africa. Known as Thermal Ground Support Base, the 2,553-acre facility had two 5,000 ft. runways. Between March 1943, and May 1944, the Army attached several liaison and tactical reconnaissance squadrons to the airfield. NAF Thermal had been inactive for six months when the Navy requested permission to occupy the base on December 2, 1944. Things were done quickly in those days and the Army gave verbal approval five days later. ACORN 29 and CASU(F)-11 arrived aboard on December 7, and began readying the station. On December 12, Commanding General, 4th Air Force gave the Navy official authorization to take over the airfield with the stipulation that the Army could reoccupy with 30-days' notice."[2]

Naval Air Facility Thermal, California. Army Photo.

NAF Thermal could house 3,456 enlisted personnel and 224 officers. It was served by a 165-bed station hospital.

On 23 December 1944 Lt. Cmdr. E. F. Zimmerman assumed command of CASU(F)-11 at NAF Thermal. He was the seventh Commanding Officer.

"Located in the Coachella Valley 150 feet below sea level, the place was named Thermal for a reason. Daily summer temperatures reached 120F in the shade soaring much higher on the concrete ramp. Conducting training here was not easy and summer flight operations took place from 0300 to 1300. In the heat of the day, the ground crews simply could not service the aircraft. At those temperatures, just touching hot aluminum would blister the skin."[3]

M.L. Shettle continued, "The base's facilities were in rather poor condition. During the first few months of the Navy's occupancy, the ACORN and Seabees made extensive improvements. Spread over four miles of desert, the usual Army tarpaper shacks were repaired and brought up to 'Navy standards.' The runways and taxiways had to be repaired and additional aircraft parking ramps installed. The Navy leased a recreation center, 25 miles from the base, with a swimming pool and dance hall for enlisted men. In addition, a local citizen supplied a house and swimming pool at a nearby ranch to the Navy. The house became the Commanding Officer's residence. Officers and their wives were allowed to use the pool. The ACORN, CASU, and Seabee training program ended on April 20, 1945, after ten such units had passed through the station."[4]

For anyone who has visited NAF Thermal, the question has to be asked – why would you isolate a CASU in the desert. The goal of CASU(F)-11 was to become experts in the repair and maintenance of the Privateer aircraft. NAF Thermal offered blue skies and unlimited airspace for the Privateer. The plane was built in San Diego, and NAF Thermal was an easy flight from there. It was also easy to transport personnel back and forth from San Diego to NAF Thermal for training at the factory. However, what the skies offered, the land did not. The area was without vegetation and the ground consisted of a very fine dust that permeated the buildings, engines, and shops. New personnel arrived daily with as many as 200 arriving monthly. NAF Thermal offered land and a miserable place to stash these men until it was time to return to Port Hueneme.

On 28 December 1944 CASU-14 arrived at NAF Thermal for training. From their War Diary;

At the time of our arrival, the facilities had been in use for about three weeks for the purpose of training ACORN-29, CASU(F)-11, and detachments of Air Group 98 - a replacement group part of an H-9 component was turned over by AATD for the use of the ACORN and CASU to maintain aircraft. The component was pitifully deficient in most of the essentials. Transportation, critically needed in such a spread out area, was limited. Maintenance equipment and tools were so short that many men were forced to stand idle while waiting for a certain tool. Furthermore, in the wild rush to get operating at once, the CASU Material Officer had broken out tools without establishing any orderly system for control of issue, and the available supply lessened steadily until this command took over and established positive control over tools.

We were informed upon arrival at NAF Thermal that our 'A' component was to acquire CASU(F)-11's 'F' components, in so much as CASU(F)-11's mission had changed from carrier type maintenance to multi-engine work. The transfer of five hundred men and eight officers to be accomplished on 1 January 1945 after which CASU(F)-11's 'A' component was to depart for Port Hueneme for staging [in preparation for going overseas to Okinawa, Japan].

It seems worthwhile at this point to discuss somewhat the history of this 'F' component we were to acquire. Most of the officers and men had been together on the west coast for several months supposedly training for shipment overseas. They had been attached to CASU-52 which had had its mission changed, and were then transferred to CASU(F)-11. Long months of uncomfortable and tiresome duty in the desert working on class 'D' aircraft, far from liberty towns, and few opportunities for leave in view of ever 'imminent' shipping orders had morale down very low. Shortly after arriving at NAF Thermal, the Commanding Officer of CASU(F)-11 committed suicide which also added to the general feeling of dissatisfaction and unrest. [Lt. Cmdr. Zimmerman was the replacement for this officer.] At the time of the proposed transfer, this component was long overdue for sea duty and might best have been broken up entirely."[5]

In early January 1945 the new CASU(F)-11 with 48 officers and 848 enlisted men moved back to AATD Port Hueneme. This location had classrooms to continue their training. Immediately adjacent to the AATD was NAS Point Magu, an operational airfield that provided opportunities for needed, hands-on, training. CASU(F)-11 was almost ready to go overseas for a second time.

Part Five - Okinawa, Japan

Late in 1944 orders were drafted for the invasion of Okinawa and those orders were carried out in the spring of 1945. This directed the return of CASU(F)-11 to the western Pacific and Guadalcanal. It was just seven months prior that CASU-11 with its 25 officers and 525 enlisted men had left Guadalcanal for the continental U.S. It was now reorganized as CASU(F)-11 with 48 officers and 848 enlisted men trained to repair and maintain the Privateer airplane.

The amphibious landing plan called for the CASU(F)-11 to split into a small, eleven man, Advanced Echelon (ADVON) and a larger, 885 man, rear echelon party. The CASU(F)-11 ADVON included Lt. Cmdr. Zimmerman, Lt.j.g. Howard Gifford, Lt.j.g. Glenn Johnson, Radio Electrician William Peer, Chief Yeoman Frank Cooper, Lt. Dan Gary, Medical Corps, Lt. Edward Lewis, Chief Aviation Electrician's Mate El Roy Chambers, Aviation Machinist Mate First Class Altus Beck, Lt. Carl Anderson, and Lt. Erdie Eubanks. On 12 February 1945, they boarded the USS Cepheus (AKA–18) in Honolulu Harbor. Even though CASU(F)-11 was temporarily separated Zimmerman continued as the Commanding Officer. On 26 February 1945 the Cepheus departed Honolulu for Operation Iceberg, the amphibious landing on Okinawa.

March 1945

CINCPAC *Operations in Pacific Ocean Areas* report March 1945: On 1 March the Fast Carrier Task Force was engaged in a strike on the Okinawa area, primarily for photographic coverage [for pre-invasion intelligence on enemy locations].

On 18 and 19 March, [Carrier Aircraft] Task Force 58 struck airfields and installations on the Japanese home islands, and enemy fleet units located in their harbors. Later in the month, the same task force conducted air strikes and bombardments of the Okinawa area in preparation for major landings scheduled for 1 April 1945.

For the CASU(F)-11 ADVON party it was an ironic coincidence that one of their first stops on 8 March 1945 was Guadalcanal. This time Guadalcanal was only a gathering place.

On 15 March 1945, the Cepheus, with the CASU(F)-11 ADVON on board departed Guadalcanal to join the Joint Expeditionary Force at Ulithi in the Carolina Islands. They arrived six days later. On 27 March 1945, the Cepheus departed Ulithi as part of Task Unit 53.2.1 of the Transport Group Baker in the Northern Attack Force for Okinawa. Transport Group Baker was part of an assemblage of "four hundred and thirty assault transports and landing ships"[1] that put men onshore on Okinawa. The remainder of CASU(F)-11 was onboard an unknown ship that would arrive off Okinawa weeks after the advance party. This scheduled gap would prove to be a significant mistake.

Why were the Allied forces, including CASU(F)-11, headed for Okinawa? E. B. Potter and Chester W. Nimitz explained this action in their seminal text, *Sea Power*, "Shortly after the American conquest of Saipan [in the Marianas.], Admiral Spruance had suggested the capture of Okinawa, 350 miles southwest of Japan. Okinawa in American hands, he pointed out, would provide airfields to supplement the bomber bases about to be established in the Marianas. The Joint Chiefs of Staff at that time rejected Spruance's suggestion, for plans were being drawn up for an invasion of Formosa [Taiwan]. The plan for Formosa however was canceled by the decision to invade the Philippines because there were not enough troops in the Pacific theater to occupy both. But whereas Luzon was as good as Formosa as a base for interdicting Japanese communications with the Southern Resources Area and for staging an invasion of the home islands, it was too far away from Japan to provide airfields for effectively bombing Japanese industrial centers. So the Joint Chiefs, reverting to Spruance's proposal, decided that they had enough troops to capture Okinawa."[2]

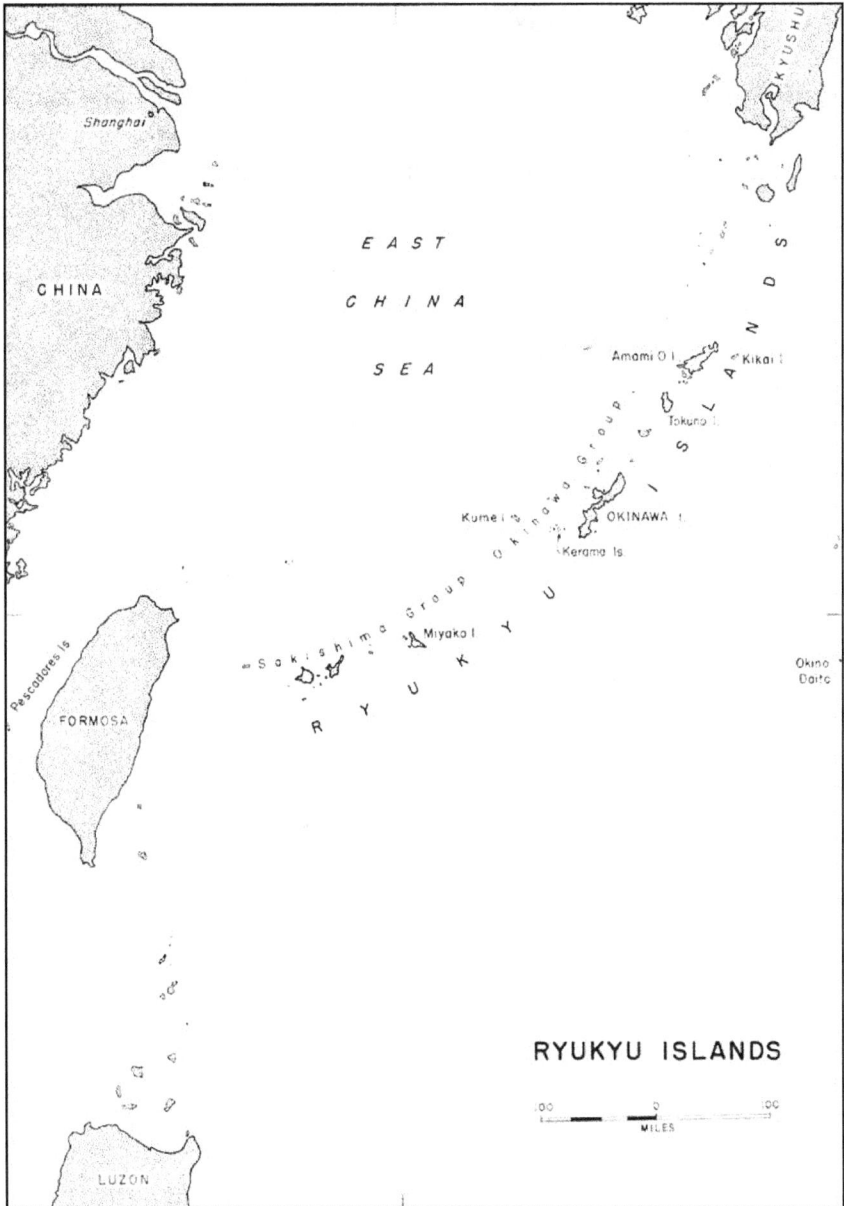

Okinawa, Formosa, China, and Japan. Army Map.

Okinawa is the fifth largest of the Japanese Islands. Okinawa means "piece of offshore rope." The island has excellent anchorages in sheltered bays, and it has a vast amount of flat land well suited for airfields. The 467-

square-mile island is approximately 70 miles long and on average seven miles wide. It is roughly 400 miles south of mainland Japan, roughly 400 miles off the coast of China, and 300 miles north of Taiwan. The southern end of the island consists of uplifted coral reef, and the geology of the northern half is marked by igneous rock. The easily eroded limestone of the south is known for having many caves. It has a humid subtropical climate.

CASU(F)-11 experienced both the monsoonal rains of spring and the torrential rainfall from the autumn typhoon passes. Typhoon season starts in June and lasts through October. Of the average 26 typhoons that form yearly in the northwest Pacific, three or four pass close enough to threaten Okinawa or at least have an impact on the weather. Locally, the farmers hope the typhoon's pass will be close enough for island rainfall but far enough away to preclude any damaging winds.

The following is from a souvenir booklet published by Naval Air Base Yonabaru during 1945, "To tens of thousands of servicemen, Okinawa, known as 'Typhoon Terrace,' 'Mud Heaven,' 'The Last Stepping Stone to Tokyo,' as well as, simply, 'The Rock,' means 'blood,' 'mud,' 'dust,' and 'spam.' It likewise means beauty of rugged hills, gnarled pines and green terraces. All about the base of these wooded hills, whose bald spots show red clay, lie a hundred baylettes of blue and emerald green waters. Along the sea walls, amidst bamboo and palm trees, one can find gorgeous specimens of Pandanuses, jasamines, amrylli, gladioli, and roses. Small villages, nestled in narrow valleys, with their red tile roofs against a background of green rice paddies and shallow streams, give an impression of a grandeur rarely found."[3]

The countryside, with steep limestone hills, umbrella-topped pines, small cultivated fields and patches of woodland, recalled those Italian landscapes seen through the windows of Tuscan and Sienese paintings."[4]

Until it began to rain! Then there was mud and running water everywhere. The island of Okinawa had mosquitos that carried malaria and fleas that produced a noticeable bite. And to add to the misery, it seemed every mud hole had standing water with maggots.

April 1945

CINCPAC *Operations in Pacific Ocean Areas* report April 1945: April operation in the Pacific Ocean Areas naturally centered around Okinawa, where the main amphibious landings were made on 1 April. This assault was the largest yet mounted in Pacific Ocean Areas, and was designed to carry our forces within easy bombing reach of the home islands of Japan. Continued naval gunfire and air support of the troops ashore were required until the island was secured in June, and defense against persistent and heavy enemy air attacks was necessary throughout the same period.

Okinawa offered numerous sites for airfields from which almost any type of aircraft could bomb Japan under fighter escort and could attack the enemy's lifelines between the homeland and the conquered territories to the south. It was a foregone conclusion that such a landing operation within 350 miles of industrial Kyushu would be strongly opposed by enemy land, air, and sea forces. The forces of the Pacific Fleet, however, had become sufficiently strong to undertake an operation this close to the enemy bases, yet 1200 miles and 2800 miles respectively from the staging areas at Saipan and Guadalcanal.

By the second of April both Army and Marine Corps forces had crossed Okinawa from its west to east coast. Enemy resistance at Yontan Airfield and the central part of the island was light.

Two enemy airfields on Okinawa (Kadena and Yontan), were captured on 1 April, and both were reported ready for emergency landing the next day. Logistically airfield sites were surveyed and developed, and shore based aviation units were set up as soon as possible. The first of these planes arrived ashore on 7 April. A land-plane search unit of Fleet Air Wing ONE (TU 50.5.5), which consisted of one VPB squadron, arrived at Yontan Airfield on 22 April to supplement the long-range seaplane searches.

The Battle of Okinawa was slightly less than 90 days, and they were horrific days. The battle began on 1 April and lasted until 21 June 1945. This was WWII's largest amphibious operation. "A British observer watching the planning for this massive operation described it as 'the most audacious and complex enterprise yet undertaken by the American amphibious forces.'"[5] The invading force consisted of over half a million Americans and no fewer than 1,457 ships.

Feifer stated in his book, *Tennozan,* "The fighting ships of this task force made up the largest assembly of combatant vessels ever gathered in naval history: over forty carriers, eighteen battleships, scores of cruisers, and almost 150 destroyers and destroyer escorts. These 318 combat vessels were roughly thrice the number the Imperial Japanese Navy had committed to the crucial Battle of Midway."[6]

On 1 April 1945, the Marines landed and it was projected that it would take until mid-April to control the Yontan Airfield. The Naval Construction Battalion assigned to ACORN 29 would then need time to repair and extend the runways and construct aircraft related buildings. The ADVON from CASU(F)-11 would accompany the ACORN and Construction Battalion and provide expertise for properly organizing aircraft maintenance work areas; for instance, engine mechanics work here and metalsmiths there. They would also service the first few aircraft that always arrived before the airfield was truly ready. Under this plan by 10 May, Yontan Airfield would be ready for air operations.

The above plan was overly pessimistic. On the first of April 1945, U.S. Marines landed along the west coast of Okinawa expecting heavy Japanese resistance. However, the 2nd Marine Division, who landed just south of the island's narrowest neck, where it is only seven miles across to the Pacific shore, found only light resistance. As the Marines worked their way to the west with the "intention to cut the island into two parts they encountered two airfields." At Yontan Airfield (also known as Yomitan Auxiliary Airfield) they found "Okinawan conscripts who had been issued uniforms but no weapons." The Marines easily overcame this underwhelming force and the airfield was quickly secured into Allied hands. Then a Zero "appeared directly overhead, almost mesmerizing the Marines below with its red rising sun. The pilot made a graceful landing, apparently with important papers from the mainland, then taxied across the field, climbed from his cockpit and walked toward an airport building before realizing that the men gathered there were the enemy. They shot him as he tried to run back to his plane;

looking down at the body, a Marine noted, 'There's always some poor bastard who doesn't get the word.'" All hands had been briefed to expect a Japanese "counterattack by a full division using parachute forces," and when this did not happen many senior staff members declared, "There must be some mistake." By nightfall, after "bulldozers cleared the runways of wrecked Japanese planes, Marine strike aircraft began landing."[7]

Yontan Airfield. Navy Photo.

During the first months on Okinawa, the rains were frequent and often torrential and swept across the entire island. In April 1945 it rained nearly every day and often at the rate of an inch an hour. Foxholes quickly filled with water. In late May at Yontan Airfield, some runways were closed as the rain could not run off fast enough to permit safe landings and takeoffs. June was equally as wet.

During April through June, the 58th Construction Battalion reported that 261 air raids were sounded at Yontan Airfield. There was a hailstorm of metal [bombs] falling from the sky. Some of it was from Japanese bombs, and some was from the Allies anti-aircraft guns on the ships in the harbor. There was such "a tremendous barrage in each raid that the harbor and airfield vicinity for miles around was prey to the never-ending rain of metal."[8]

Frank Nilson, a Sailor assigned to drive a fire truck for ACORN 29, reported, "ACORN 29 started landing on 2 April, they had to get moving as they had the gear required to get Yontan Airfield up and running. Mid-afternoon they lifted my fire truck over the side into a smaller boat for transport to the beach. There the Seabees had set up a short pier, our boat lowered a ramp onto the end of the pier and I proceeded driving the truck up the ramp to the pier. The pier was so narrow I had only a couple of inches on each side of the truck's tires. I had to really listen well to the shouted directions so I could stay on the pier, which I did, and the truck made it up on the beach, from there it was a short drive to Yontan Airfield."[9]

Over the next few days the vehicles pictured below were delivered to Yontan Airfield.

Advon CASU-11 tank trucks used for fueling aviation units at YONTAN and KADENA, OKINAWA. First trucks came ashore April 4, 1945.

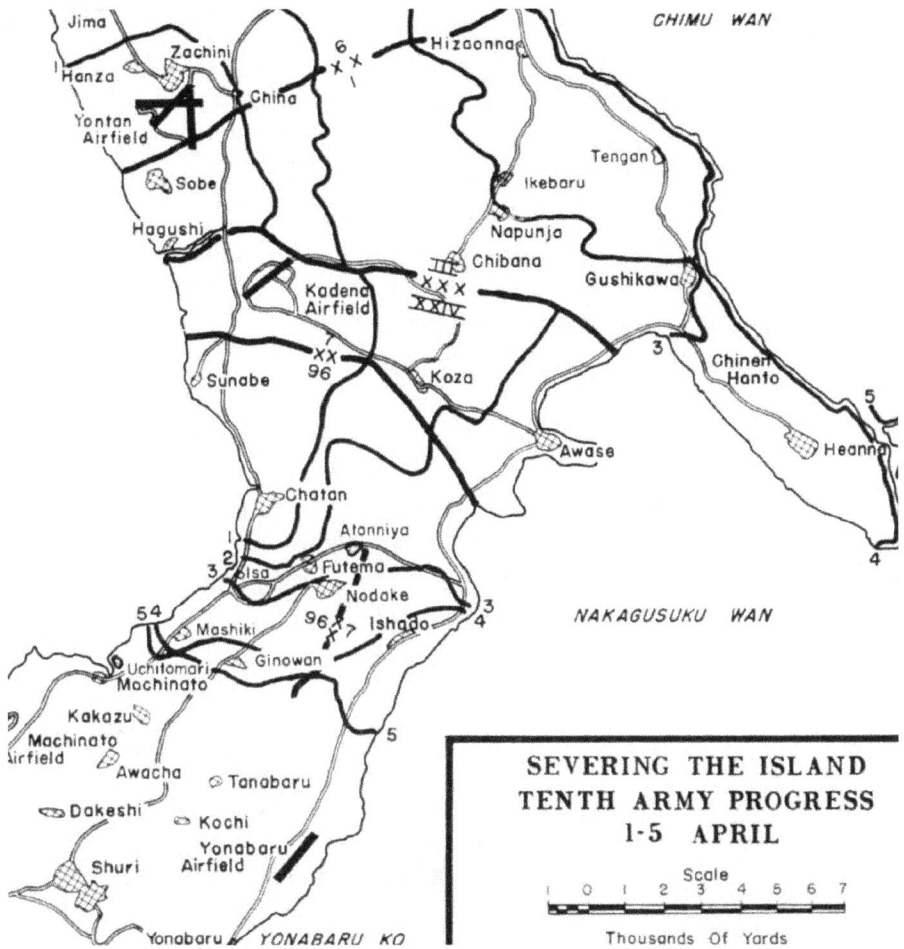

Yontan (north), Awase (projecting out into Nakagusuku Wan) and Yonabaru (south) Airfields, Okinawa. CASU(F)-11 will eventually have a presence at all three locations. Army Progress Map.

On 3 April, the CASU(F)-11 ADVON went ashore in Cepheus's landing craft. They quickly made their way 1,000 yards from the beach to the air operations control tower near the intersection of Yontan Airfield's two crossing runways.

During the Okinawa operation ACORN 29 and CASU(F)-11 were assigned to mutually support each other. According to the ComAirPacSubComFed Air Logistic Plan No. 1-45, Annex A, "ACORN

29 will support the following squadrons: (1) VPB-106 and VPB-118 which are fifteen (15) plane PB4Y-2 squadrons, (2) VD-3 which is an eight (8) plane PB4Y-1P photographic squadron and (3) VPB-133 which is a twelve (12) plane PV squadron. The CASU(F)-11 ADVON will take to the objective airfield "emergency spares for carrier type aircraft: TBM-3, F6F, FM-2 and SB2C-4. These spares covered only those items considered necessary to provide strip-side service."[10] Additionally, "An initial thirty (30) day allowance of emergency spares for four to six PB4Y-2's staging are included with the carrier type spares listed above and are to go forward with CASU(F)-11 Advon."[11]

The logistics planners significantly underestimated the quantity of aircraft that were going to land during April at the new Yontan Airfield. Subsequently, they severely underestimated the required maintenance and repairs for these aircraft. On 3 April 1945, the entire approximate 800 men of CASU(F)-11 should have landed along with the advance party. The work load far exceeded what the eleven man advance party could handle.

CASU(F)-11 requested an immediate, emergency augmentation of men. On 30 April two officers and 33 enlisted men from CASU-44 on Tinian joined the seven officers and four enlisted of the CASU(F)-11 ADVON. They remained with CASU(F)-11 at Yontan Airfield through 8 June.

From the CASU-44 War Diary, during this period these two small CASU contingents experienced, "over one hundred air-raids in less than thirty days that they were in Okinawa. On 21 April air alerts were sounded at 0000 to 0405, 1115 to 1135, 1603 to 1645, and 1932 to 2400. Three bombs killed one and wounded four enlisted personnel of Commander Naval Air Base Yontan staff in the CASU(F)-11 work area and destroyed or damaged five tents and personnel gear. One bomb made a bullseye on the CASU(F)-11 ice cream freezer also destroying the bakery."[12]

On Okinawa the men of CASU(F)-11 experienced a war unlike what the men of CASU-11 endured at Guadalcanal. The Japanese were throwing everything that they had against the Allied forces near and on Okinawa. Rumors circulated that manned Japanese torpedoes filled with explosives had been released into the Navy fleet anchored just offshore. These were

"Kaitens" and luckily only a few were employed by the Japanese off Okinawa.

During the first few days the men on Yontan Airfield would see Japanese Kamikaze planes streaking downward from the sky and attempting to sink the nearby Allied fleet. Everyone in CASU(F)-11 had heard about Kamikaze suicide aircraft. However, on 6 April 1945 they had their first experience with these raids and no one expected the 350 Japanese Kamikazes plus the 300 fighters and bombers that attacked the Allied shipping and ground forces ashore that day. Over the next 70 days the Japanese flew hundreds of suicide planes towards Okinawan targets, Allied airfields and shipping lanes.

From pilots and other U. S. military personnel came stories of the Marines finding 250 "Shinyo," high speed suicide boats hidden on Kerama Retto. This is a group of small islands fifteen miles west of Okinawa. The boats with one seat for the driver were loaded with 1,000 pounds of high explosives.

It was at Yontan Airfield that the American forces first saw an Ohka. The Allied name for this little horror was "Baka." "It was a manned glider bomb with three rockets as boosters and a warhead carrying 2,645 pounds of tri-nitro-anisol. The Baka would arrive in the combat area slung under the belly of a two-engined bomber. Its pilot had communication with the bomber's pilot through an umbilical wire, who released him near the target. The suicide pilot had to pull out of a vertical dive into a glide toward the victim. Then, if necessary, he would increase speed to over 500 knots by the use of the rockets. The small size and tremendous speed of the baka made it the worst threat to our ships operating off Okinawa."[13]

In addition to all of the above, on the morning of 6 April, the Allies learned that the mighty Japanese battleship Yamoto had sailed on a one way trip to Okinawa. After the war the suicidal nature of this effort was verified "by the fact that Yamato was only given enough fuel for a one-way trip."[14] The Yamoto planned to cut a huge swath with her guns through any attacking Allied ships and then drive onto the beach and become an unsinkable fortress. This beach could have been the one located only a yards from Yontan Airfield and CASU(F)-11. It was the heaviest and most powerfully armed battleship ever constructed, and it warranted considerable fear.

Fortunately, at dawn on 7 April a U.S. submarine discovered and quickly reported Yamato's position. Multiple carrier battle groups in the area launched aircraft and four waves of aircraft mercilessly bombed the Yamato. By 1400 the battleship had a 70 degree list and was almost stationary in the water. Twenty three minutes later the ship rolled over and one of the two bow magazines detonated in a tremendous explosion and the ship sank.

During the attack an Avenger TBM Torpedo Bomber "was hit during pull out after attacking the YAMATO. The entire crew parachuted, but only the pilot survived. He noted that during most of the time he was in the water the battleship was dead in the water and smoking heavily."[15]

Two PBM Mariner flying boats were in the area shadowing the YAMATO and providing location reports when they observed the TBM shot down and crash landing near many Japanese survivors. Because of the possibility of the pilot being captured, plus the smooth condition of the sea, one of the PBMs made an open sea landing and rescued the TBM pilot. Meanwhile the other PBM circled to draw the gun fire from the nearby ships away from the rescue."[16] As we learned from CASU-11 operations on Guadalcanal the Japanese rarely made an effort like this to rescue one of their own pilots.

On 12 April 1945 shocking and sad news from home arrived at Yontan Airfield. President Franklin D. Roosevelt had passed away. For most of the men in CASU(F)-11 President Roosevelt, "was the only president they had ever known in their lives and they took the loss personally. Many wept.[17]"

Later in the day, several enemy shells apparently fired about 12,000 yards behind the front lines, struck Yontan Airfield. Fortunately, the damage from these attacks was negligible.

On 18 April 1945, another well-known American hero was lost: writer, Ernie Pyle. The day before, Pyle had landed with the Army's 305th Infantry Regiment on Ie Shima, a small island near Okinawa. "The next morning, riding with a battalion commander, suddenly a Japanese machine gun opened up, and the driver with his two passengers dived into a ditch. After the machine gun fell silent, the commander and Pyle raised their heads, and the gun chattered again. Pyle slumped back into the ditch. Bullets had entered his forehead just below his helmet."[18] Similar to Pappy Boyington and Charles Lindbergh, the men of CASU(F)-11 only knew Pyle from a distance; however, they had all read his news articles and viewed his cartoons. His name, like Pappy Boyington and Charles Lindbergh, was talked about by CASU(F)-11 veterans for years after the war ended.

With the arrival of the Privateer on Okinawa during the early months of 1945 the Navy was quickly convinced that this airplane could conduct the required multi-hour, longer range, search and armed reconnaissance missions against enemy surface forces and aircraft. Typical missions covered a thousand miles for each search sector. These included security patrols around Okinawa. Missions also included weather reconnaissance and special courier flights from Okinawa to Shanghai, China. The Navy also realized that missions with "combat pairs," two Privateers flying together, had more guns and created a measurable advantage against inbound enemy fighters.

In an April 1945 after-action report by Commander Naval Air Base Okinawa, it was reported that "Maintenance of aircraft in spite of limited personnel was satisfactory during each stage of the operation. The CASU(F)-11 (ADVON) was tailored to support four to six Privateers during the initial phase [of the landing on Okinawa], instead, [they] supported initially six VPB aircraft which were assigned to operate from Yontan Airfield, then a squadron of fifteen VPB aircraft, and subsequently, one and one half squadrons, all ahead of proper regular supporting ACORN and CASU personnel and equipment. During the period 8 to 30 April 45, CASU(F)-11 and the flight section of Air Base Command effected repair services on 74 carrier type aircraft [fighters and bombers] at Yontan Airfield."[19] The advance party of CASU(F)-11 at Yontan Airfield desperately needed the arrival of the remaining 872 men of CASU(F)-11.

The next picture is a B-29 that CASU(F)-11 repaired and sent back to its home base.

B-29 returning from TOKYO raid with two engines out made an emergency landing on YONTAN Field, OKINAWA, on April 19th. ATC plane flew in replacement engines. Repair crewmen and CASU-11 personnel made repairs which enabled plane to return to home base.

May 1945

CINCPAC *Operations in Pacific Ocean Areas* report May 1945: During May nearly all efforts of Central Pacific surface and air forces were directed toward continued support of the Okinawa campaign, which lasted from 26 March to 21 June. Heavy ground fighting on Okinawa, supported by naval gunfire and air strikes, continued throughout May, with a heavy toll on our ships sunk or damaged by persistent suicide attacks launched from Japanese bases. The effect of attrition on enemy air forces was, however, becoming evident in the fact that such attacks were markedly lighter and less frequent than had been the case during April.

The activities of Fleet Air Wings One and Eighteen, included long range searches, neutralization missions, reconnaissance, offensive anti-shipping sweeps, and area coverage on a large scale. Their forces consisted of seven squadrons of 'Mariners' and 'Coronados' operating from Kerama Retto, southeast of Okinawa, and "Privateers" based at Yontan Airfield, Okinawa. In May alone, these units sank 94,415 tons of enemy shipping (compared to 6160 in April), and damaged 91,020 tons (as compared with 17,412 in April). Action by the enemy against Yontan Airfield hampered operations; from 8 through 10 May, 'Privateers' were unable to launch flights.

The heaviest enemy air attacks during the period, however, came on the 4th, 11th, 24th, 25th, and 27th of May. Early on the morning of 4 May, approximately 60 enemy planes approached the objective area, dropping bombs on the Yontan Airfield runways, which, however, remained operational. Around 0700 on the same day, approximately 70 planes, in a series of 14 raids, attacked the objective area, sinking Luce and Morrison (DDs), 1 LCS and 2 LSMs by suicide crashes, and damaging Birmingham (CL) and five destroyer types. A third attack, by 30 planes, at 1900 on the same day, resulted in damage by suicide attacks to Sangamon (CVE) and Gwin (DM). During the day, 137 enemy planes were destroyed in the objective area, including 12 suicide hits.

On 11 May, when the Tenth Army (Japan) launched a coordinated attack on the entire front, there was heavy enemy air attacks against shipping in the Okinawa Area. From 0100 to 0500 and from 0800 to 0930, approximately 110 enemy planes were in the area. Hadley and Evans (DDs), 1 AKA, and 1 LCS received suicide hits, but 93 enemy planes were destroyed in the two attacks. Task Force 58 was also under attack by enemy aircraft during the day, and shot down 35 planes. Bunker Hill (CV), flagship of CTF 58, received serious damage from a suicide hit, and CTF 58 shifted his flag to Enterprise (CV).

Beginning at dusk on the 24th, and continuing until 0400 on the 25th, and also from 0730 until 1130 on the 25th of May, the enemy made concentrated and persistent attacks on units ashore and afloat. The Japanese employed an innovation in their air attacks, when five enemy twin-engine planes attempted to land at 2225 on the 24th on Yontan Airfield. AA accounted for four of them, but one plane landed, wheels up, on Yontan Airfield, discharging about 10 enemy soldiers who attacked the transport line with demolition charges, destroying 7 planes, and damaging 25. All of the enemy were killed, but they succeeded in putting the field out of commission until early the next morning. During the attacks on the 24th and 25th, 195 enemy planes were destroyed, including

11 suicide crashes. During these attacks, nine ships were damaged, and one LSM and Bates (APD) were sunk by suicide crashes.

From 2000 on the 27th, and continuing until 0930 on the 28th, large scale air attacks were in progress over the objective area. A total of 123 enemy planes were destroyed, including 11 suicide crashes. During the same period, seven ships were damaged; and one LCS and Drexler (DD) were sunk by enemy suicide attacks.

Even though they were seriously undermanned at Yontan Airfield, the Advance Echelon of CASU(F)-11 was noticeably productive. In a 2 May letter Commander Task Force Group [CTG] 58.4 noted "an increasing number of aircraft from carriers of the group have been landing at Yontan Airfield because of anti-aircraft fire damage or other failures." The letter further stated that in getting their aircraft repaired or replaced, pilots should work with CASU(F)-11. The letter continued, "the proposed method of turning aircraft over to CASU is the normal one. It apparently has not been used by our pilots because it was not expected that the CASU would be ready at such an advanced date. The fact that it is ready is highly commendable, and full advantages should be taken of it to obtain further use from aircraft which cannot be landed on the carriers." CTG 58.4 ended his letter by recommending, "the complement of CASU ELEVEN should be increased more rapidly than planned particularly as CASU ELEVEN is not charged primarily with the responsibility of handling carrier aircraft, but is also responsible for many land-based aircraft."[20]

The following is the log for a typical day in May 1945. "Air alerts at Yontan Airfield were observed from 0140 to 0155, 0203 to 0207, 0245 to 0445, 1050 to 1115, 1910 to 1915, and 1930 to 1950. At Yontan Airfield between 0300 and 0400 the following bombs were dropped by the enemy aircraft:

Ten (10) 500# fragmentation bombs north of the N/S runway.
One (1) 100# fragmentation bomb near the control tower.
Two (2) 100# fragmentation bombs in the CASU(F)-11 area.
Two (2) daisy cutters in the MAG-31 area.
Two (2) phosphorous incendiaries in the MAG-31 area.

Two (2) large bombs causing craters six (6) feet deep and ten (10) feet in diameter on the field east of the N/S runway.

Two (2) delayed fuse bombs in CASU(F)-11 area which exploded at 1130 and 1900.

Six (6) 100# fragmentation bombs in the Corps Evaluation Hospital #3 area, just south of the fields.

[A 'daisy cutter' was a bomb that exploded just before it buried itself in the ground. Many bits and pieces of shrapnel flew horizontally in all directions. The bomb was used to damage many parked aircraft or kill many men in a tent area.]

"The damage resulting at Yontan Airfield was as follows:

"CASU(F)-11 equipment damaged was: One (1) weapons carrier; two (2) gasoline trucks; one (1) Cletrac tractor, and two (2) engine test stands damaged by shrapnel. No injury to personnel of CASU(F)-11 resulted.

Cletrac Tractor. Navy Photo.

"Corps Evaluation Hospital #3 had several tents riddled by shrapnel and one (1) bombproof damaged. Personnel injured included twelve (12) killed and thirty (30) men wounded."[21]

The miserable wet weather of April became worse in May. "And wet days they were. Rains set in on the 16th, and from that day to the end of the month

193

there was a precipitation of almost 21 inches. The resulting mixture of rain water and Okinawa soil produced such mud as none of the Americans present had seen or imagined."[22] Bulldozers and trucks were almost completely useless.

Wet feet were problematic for nearly everyone. Sledge in his book, *With the Old Breed at Peleliu and Okinawa,* reported, "During a period of about fourteen or fifteen days, as near as I can calculate the time (from 21 May to 5 June), my feet and those of my buddies were soaking wet, and our boondockers [boots] were caked with sticky mud. It was hard to get your feet dry and almost impossible to clean the mud off your boots. Consequently most men's feet were in bad shape. My feet were sore, and it hurt to walk or run. The insides of my boondockers gave me the sensation of being slimy when I wiggled my toes to try and warm my feet with increased circulation. The repulsive sensation of slippery, slimy feet grew worse each day. [Unfortunately, this was not a situation where one could buy a new pair of boondockers every few days.]

The almost constant rain also caused the skin on the hands to soften with a strange shrunken and wrinkled appearance. Sores developed on the knuckles,"[23] it made it very difficult to handle tools and work outside on the aircraft. Everything was wet or damp. Shoes, socks, clothing, bedding, nothing ever got dry. Jungle rot, just like the men experienced in Guadalcanal, reappeared. Starting with wet feet, then the crotch and underarms, and finally getting to the skin area under your belt. The old treatment, bottles of purple potassium permanganate, was again distributed to all personnel.

Leckie described the wet conditions, "Nothing could stand against it; a letter from home in the sodden pocket of a [Sailor], GI or Marine had to be read and re-read and memorized before the ink ran and it fell apart in less than a week; a pair of socks lasted no longer; and a pack of cigarettes became watery and uninflammable unless smoked the same day, or else, along with matches, they were kept dry within a contraceptive inside a helmet liner. Pocketknife blades rusted together, and watches recorded the period of their own decay. Rain made garbage of the food; pencils swelled into useless pulp; fountain pens became clogged with watery ink, and had to be slung upside down to keep the raindrops from fouling their bores."[24]

Flies were everywhere. "If you put down a cup of coffee for one second, you could hardly see it when you looked again. The surface of the coffee was covered solid by huge black insects. It was believed that flies making trips from battle field corpses to filled coffee cups in open air kitchens miles away helped spread dysentery, a disease practically every person in Okinawa had more than once during their time on island."[25]

Moving from one place to another was a struggle. When it was dry weather vehicles in motion would produce blinding clouds of red dust, and then when it rained the resulting mud stuck to tires. Jeeps, trucks and bulldozers all needed men with shovels to dig them out. When a plane would crash, of course, it would not stop on the runway and be easy to tow away. No, it would always end up in the mud.

On 8 May 1945, CASU(F)-11 received the following commendation from the Commander Naval Air Bases, Navy No. 3256, "You are commended for the excellent work of CASU(F)-11 in connection with repair of damaged carrier aircraft during the first month of operations. This initiative and industry on the part of your command has resulted in much assistance to the Fast Carrier Group."[26]

On 9 May between 2155 and 2210, enemy aircraft dropped an estimated 8 bombs that resulted in 6 holes in the NE-SW runway. By 2330, the runway was again operational. On 10 May from 0232 to 0440 the enemy flew at altitudes of 15,000-18,000 feet over Yontan Airfield. During this period Allied anti-aircraft guns fired continually.

On 10 May Patrol Bombing Squadron 109 arrived at Yontan Airfield. VPB 109 was the first of three squadrons equipped with the new Special Weapons Ordnance Device (SWOD) Mark 9 "Bat." VPB-123 and VPB-124 soon followed. Upon their arrival all 24 Bat maintenance personnel were assigned to CASU(F)-11 for temporary duty.

The Bat was a new weapon system and required specifically trained technicians. "The squadron was based on CASU(F)-11, which had only an advanced echelon ashore and was seriously handicapped. There were only nine officers and 57 enlisted men. CASU(F)-11 personnel were trained in PB4Y maintenance, but facilities were available for servicing only a squadron echelon [three to six aircraft] instead of the two squadrons [18 to 24 aircraft] onboard.[27]

The U.S. Navy's ASM-N-2 "Bat" glide bomb was one of the world's earliest fully-automatic, target-seeking "smart" weapon system. Essentially it was a plywood glider shell wrapped around a 1,000-pound general purpose bomb. The Bat had a search and homing radar with a range of approximately 25 nautical miles (nm) and could lock on a target at about 10 nm. Thirty-five test drops were made under combat conditions in 1945, resulting in a total of four direct hits. Even with this low number of successes during testing it was decided to send the Bat off to war.

An ASM-N-2 Bat guided bomb on a Consolidated PB4Y-2 Privateer. Navy Photo.

Once in the Western Pacific, it was found that the Bat was very sensitive to corrosion. In retrospect the humid climate of the islands in the Pacific presented many challenges. The VPB 109 War Diary noted, "The Bat men struggled with overwhelming difficulties. The Bats had been flown through two typhoons with subsequent warping and deterioration. They had been assembled and disassembled numerous times, stacked in the hot sun and dust of the Philippines and the mud and mold of Okinawa, unprotected and with no slight gentleness. No test equipment was available, and it is a tribute to the perseverance and untiring energy of the Bat men that the weapon managed even negative results."[28]

On May 27 while off the coast of Korea, Lt. Leo Kennedy flew a VPB-109 Privateer that crippled the 970-ton Japanese escort ship Aguni with a Bat from 20 miles away. Subsequently, Kennedy destroyed a 2,000-ton freighter and three smaller freighters with a regular bomb. For this 90 minutes of action, Lt. Kennedy was awarded the Navy Cross. Lt. Kennedy may have been the first pilot to have received a U.S. decoration based on his use of an automatic homing missile.

During May 1945 CASU(F)-11 and VPB 109 at Yontan Airfield experienced very dangerous conditions. "Foxholes were a necessity at Yontan Airfield. Nightly at sunset came the first alert, and actual raids were frequent. The enemy was largely accurate in his bombing; though, even from altitude, bombs invariably struck the field and rarely struck near personnel areas. Everyday kamikaze planes were observed striking nearby ships or shore installations. Personnel endured loss of sleep and time spent in foxholes. Operationally, the nightly alerts forced all patrol flights into daylight hours, for the planes could not be serviced or gassed or takeoff during an alert. Similarly, all flights returned before dark to avoid mix-ups with the numerous bogies and the fleet's understandable propensity to fire at anything with wings after dark."[29]

"In his tent at Yontan Airfield, the pilot, Samuel Hynes, identified two kinds of fear from shelling. The first, constant and inescapable even for personnel in the relatively secure rear, was not knowing the moment when a shell with your name on it would land. But that was mild compared with the fear during the actual shelling, when the bravest men couldn't stop trembling."[30]

"At 0324, on 16 May, enemy aircraft dropped 12 bombs on the east side of the NS runway. One transient fighter aircraft was hit and burned."[31]

Finally on or about 17 May 1945 the 885 remaining men of CASU(F)-11 arrived at Yontan Airfield. The date of arrival is approximate because research did not reveal the exact date. The suggested date was based upon two facts. First, the Navy Awards Manual listed 6 April 1945 as the date the CASU(F)-11 Advance Echelon was eligible for the Asiatic-Pacific Campaign Medal and it used 29 May 1945 as the start date for the CASU(F)-11 "rear" echelon's eligibility. Secondly, Mr. Greg Bingham, private photo collector, provided the photo below from 24 May 1945 of five men from CASU(F)-11. The names of these men Deeck, Gibson, Pally, Grace and Sykes were written on the back of the photo and they were not members of CASU(F)-11's Advanced Echelon. They had to be part of the rear echelon and the rear echelon group was on Okinawa at least a couple of days prior to the evening events of 24 May 1945.

CASU(F)-11 Rear Echelon members with souvenirs. From Bingham.

"On 18 May at 2215 bombs hit Yontan Airfield and started a small oil dump fire. The bombs also hit a PB4Y-2, Privateer, while it was being maintained by CASU(F)-11. The plane, subsequently, exploded. The NE-SW runway was cluttered with its wreckage. The N-S runway was damaged with 3 large holes; 16 aircraft were hit by fragmentation bombs, 3 suffered major damage."[32]

By mid-May 1945, the Japanese General Ushijima realized that, without help, Okinawa was going to be lost. Feifer, in his book Tennozan, stated that "the general came as close to imploring (headquarters for help) as any Japanese general could - and won a promise of commando-type raids." These would be added to the already ongoing aerial bombardment. "[The raid] was actually mounted two days later on Yomitan (Yontan) Airfield. That field, taken on L-day, now teemed with American planes refueling and loading armament for more strikes."[33]

On the evening of 24 May 1945, "the most ferocious display of antiaircraft power yet seen in the Pacific broke up a daring airborne attack on Yontan and Kadena Airfields. It was an unusually clear night, and there were thousands of witnesses to this small savage setback that the suicide spirit was able to inflict on the Americans. Perhaps twenty twin-engined [Sally] bombers came gliding through a fiery lacework woven by American antiaircraft gunners. Eleven of them fell in flames. The rest, except one, fled."[34] The next photo shows the amazing amount of antiaircraft gunfire put into the air that night.

Anti-aircraft gunfire protecting Yontan Airfield. Navy Photo.

The number of aircraft involved in this attack varies from one witness to the next. However, all agree that one plane crash landed in flames in the middle of CASU(F)-11's work area with 12 Japanese commandos and two pilots. Not all survived the crash, however "at least eight heavily armed Japanese rushed out of the plane and began tossing grenades and incendiaries into American aircraft parked along the runway. They destroyed two Corsairs, four C-54 transports, and one Privateer. Twenty-six other planes, one Liberator bomber, three Hellcats, and 22 Corsairs were damaged. In addition to these thirty-three planes destroyed and damaged, two 600-drum fuel dumps containing 70,000 gallons of gasoline were ignited and destroyed by the Japanese"[35]

Mitsubishi Ki-21 (Sally) two engine bomber. Army photo.

Naval Air Base Okinawa, in their May 1945 War Diary, reported the attack as follows, "Air alerts on the night of May 24-25 started 2020 24 May. At 2105 AA fire was noted both north and east of Yontan Airfield. At 2110 six (6) bombs and at 2204 four (4) bombs of undetermined size were dropped northeast of the airfield with no damage resulting. Single enemy aircraft, type unknown strafed Yontan Airfield at 2210 and again at 2220 causing no damage. This was followed at 2225 by three (3) Sallys (Allied name for Japanese bomber.) attempting belly landings Yontan Airfield. All three crashed and burned due to pilot trouble or were shot down by low heavy concentrated AA. At 2230 two (2) more Sallys attempted belly landings on Yontan Airfield, one indicated as B1 was successful and one at B2 crashed and burned as a result of pilot undershooting runway or AA fire. Location of these planes are indicated as follows:

A1 – 'Sally' ½ mile NE of North end of Yontan Airfield.
A2 – 'Sally' ½ mile E of North end of Yontan Airfield.
A3 – 'Sally' 100 feet east center of middle of NE/SW runway, Yontan Airfield.
B2 – 'Sally' East edge at north end N/S runway Yontan Airfield.

"Tower personnel passed the word 'Jap gliders are landing on Yontan field' after seeing airplane B1 slide to a stop and about twelve (12) disembark and scatter to nearby aircraft parked in the tower area and in the E/W runway. Tower duty officer and two assistants put a spot light on enemy disembarking, while field crash truck and radio jeep personnel dashed along NE/SW and E/W runways in vicinity of scattering enemy troops. Enemy planes at A1, A2, A3 and B2 were burning at this time. Within five minutes additional fires were started by airborne Japanese from enemy Sally indicated B1."[36]

CASU(F)-11/44 personnel were working in the tower, on the crash truck, and on various aircraft parked around the tower. "CASU personnel killed five (5) of the attacking group but none of the CASU members suffered any injuries."[37]

Two (2) Japs were killed by rifle fire from Yontan Airfield crash truck. The tower duty officer killed one Jap but was wounded in the stomach by Japanese fire. Several Jap troopers were busy throwing hand grenades at the control tower which wounded two assistant operators. Four (4) men in the crash truck and one (1) man in the radio jeep were wounded.

Investigation revealed that five Jap 'Sallys' reached the Yontan Airfield area. All of these planes had no armament other than automatic weapons. It is estimated that each "Sally" carried gasoline for a one way trip to Okinawa to prevent the planes catching fire on landing.

The "Sally" shot down by AA fire adjacent to Commander NAB camp at A3 entombed the Marines AA Battery personnel of eight (8). Construction Battalion Maintenance Unit No. 617 personnel and others fought fires amidst exploding incendiaries and grenades to rescue six (6) wounded and two (2) dead Marines of the Air Defense Command.

Fifty-eight (58) enemy were killed as a result of crashes or flak. By 1800, 25 May 1945 seventy-one (71) Jap dead were counted in Yontan Airfield area and buried by Construction Battalion Maintenance Unit No. 617. During the night of 25-26 May a naked Jap wrapped in a blanket was caught alive in VMF-441 area, south side of Yontan Airfield and turned over to the Military Government.

Best estimates indicated that there were thirteen (13) to fifteen (15) Japanese including the pilot and crew in each airplane. Positive evidence shows all of the enemy to be of good military physique, well fed, clothed and equipped for their obvious mission.[38]

There is an interesting and informative movie about this multi-aircraft Kamikaze raid on Yontan Airfield. It can be found at: https://www.youtube.com/watch?v=zR8usETx6Qg.

VPB-109 reported, "on the last day of the month [May], the squadron was relieved at Okinawa by VPB-123 and VPB-109 was ordered by the Wing Commander to retire to Tinian for rest and repair. The maintenance problems had exceeded the CASU's abilities, and only two planes remained in a state of combat readiness. Despite harassment and primitive conditions, the men had stood up better than the machines."[39]

June 1945

CINCPAC *Operations in Pacific Ocean Areas* report June 1945: The Okinawa campaign, the largest yet undertaken by Pacific Ocean Area forces, was completed on 21 June, when the island was declared secure after nearly three months of heavy fighting by naval, air, and ground forces. Logistic forces proceeded with the development of Okinawa as a large base to support future operations against the Japanese home islands. A severe typhoon on 5 June, which some of the forces supporting the Okinawa operations were unable to avoid, was responsible for nearly half of the damage sustained during June by ships of the Third Fleet.

Land-based air forces of the Pacific Ocean Areas steadily increased the scale of their attacks against the Japanese homeland and shipping, encountering a minimum of opposition from Japanese aircraft. It was evident that the Japanese air force had been withdrawn from the battle areas, in an effort to muster enough strength to oppose the final assault on their homeland.

With the mid-May arrival of the rear echelon with 885 enlisted personnel, CASU(F)-11 was now at its full complement. The commanding officer could release the 44 members of the CASU-44 augmentation group. On 8 June 1945 they returned to Tinian.

On 10 June 1945, while VPB-109 was continuing rest and repair on Tinian, they transferred thirty-one SWOD (Bat) personnel for temporary duty to CASU(F)-11 on Okinawa. This continued the consolidation of all deployed SWOD (Bat) personnel at CASU(F)-11. Aircraft Radio Technician Second Class John Bernard Goodman, Jr., was among the men transferred from Tinian. His nephew Mark shared the above information about the history of the SWOD (Bat) program and CASU(F)-11 on Okinawa.

Starting in early June and completed by 13 June all the personnel of the Patrol Bombing Squadron 123, including twenty-two men assigned to CASU(F)-11 for SWOD support, had reported to Yontan Airfield.

The squadron was given one day in which to orient itself and attempt to get settled. Accommodations were indeed very poor to say the least. The weather was worse; it was raining when the first increment arrived, and it continued for approximately two weeks without let-up.

For the first five days the squadron office consisted of one collapsible chair and a cruise box in a remote corner of the Landplane Detachment's tent [a small-sized edition of a "circus" tent, which leaked and sagged with the rain and wind]. CASU(F)-11 Supply generously gave us a new 14 X 14 ridge pole tent, we promoted a load of coral, sidetracked a wandering bull-dozer to level the coral, pitched the tent and established an office.

CASU(F)-11 handled the maintenance on the aircraft. In spite of the lack of equipment and materials it did a commendable job in keeping up with the availability of aircraft. ACORN 29 on which the CASU was based seemed to employ 'hoarding' tactics with its equipment. The ACORN unit itself never appeared at Yontan Airfield, it was building Yonabaru Airfield, and refused to make trucks, jeeps, caterpillars, etc., available to ease the pressure and congestion at Yontan Airfield.[40]

On 16 June 1945, the third and final Patrol Bombing Squadron (VPB-124) arrived with SWOD (Bat) repair and maintenance personnel. These twenty-three men were assigned temporary duty with CASU(F)-11. VPB 109, 123 and 124 had now all contributed personnel to this Special Weapon Ordnance Device "Bat" weapon effort centered at Yontan Airfield, Okinawa.

On 4 June 1945 at 0100 in the CASU(F)-11 Camp area, a Jap infiltrator was driven from a tent by small arms fire. He made his escape to the hills.

VPB 124 reported in their war diary that, "Maintenance, on the whole, was good, considering the conditions under which the maintaining units operated. In this connection, this command would like to acknowledge the work done by CASU (F)-44 (Tinian) and CASU(F)-11 (Okinawa)."[41]

On 21 June 1945, the Marines declared Okinawa secure from the Japanese. The Battle of Okinawa was the bloodiest and most costly campaign in the War in the Pacific. CASU(F)-11 supported it all. The Americans suffered 49,151 battle casualties. They lost 763 aircraft, 36 ships, and another 368 ships were damaged.

The Japanese, including conscripted and drafted civilians, lost 110,000 men. Regular combat and kamikaze aircraft losses was a staggering 7,800 airplanes. Among these were ten major kamikaze attacks with a loss of 1,465 airplanes; and 1,900 individual suicide sorties.

"These mass attacks were designated by the Japanese *kikusui (floating chrysanthemum)*.

Attack	Date	Number of Japanese planes
1	6-7 April	355
2	12-13 April	185
3	15-16 April	165
4	27-28 April	115
5	3-4 May	125
6	10-11 May	150
7	23-25 May	165
8	27-29 May	110
9	3-7 June	50
10	21-22 June	45
	Total	1465"[42]

The Japanese Navy lost 16 ships, and four more were seriously damaged. They lost their last great battleship Yamato. After Okinawa the Japanese Navy was completely incapable of mounting another meaningful attack.

On 29 June 1945, CASU(F)-11 received another commendation from the Commander Air Force Pacific Fleet, Subordinate Command Forward, Area. It was forwarded to the Chief of the Bureau of Aeronautics. The "Casu(F)-11 advanced echelon arrived on Yontan Airfield, Okinawa, on D plus 5 to service aircraft of Fleet Air Wing ONE. Although on numerous occasions this unit has operated under both bombing and shelling by the enemy, they have performed an excellent job in aircraft maintenance under adverse conditions. The photographs, enclosed, show typical conditions of the working area caused by several days of constant rain, and indicate the type of battle damage and repair work being done on PB4Y-2 aircraft. Since the photographs were taken, additional personnel and equipment have arrived, and maintenance facilities are being constantly improved. The ability of Casu(F)-11 to operate under adverse conditions, and to improve where necessary, extended greatly the maintenance services at the field."[43]

Unfortunately, the referenced photos are not available.

As the Battle for Okinawa reached its conclusion, the market for war souvenirs grew. CASU(F)-11 was in the center of this marketplace with pilots who had money, and troops who hitchhiked to the airfield to sell or trade their finds.

When a Japanese pilot landed at Yontan Airfield thinking it was still in Japanese hands, he found himself surrounded by U.S. Marines. The "Marines rushed to strip the pilot the moment they shot him"[44] and then went to his plane to collect other war trophies. The war relic collecting on Okinawa was different from the seemingly laid back days on Guadalcanal. As on Guadalcanal, prized items still included Japanese battle flags, especially those signed, and samurai swords. However, on Okinawa, prized items also included body parts: teeth, fingers, and ears.

July 1945

CINCPAC *Operations in Pacific Ocean Areas* report July 1945: While ground and amphibious forces of both the Central Pacific and Southwest Pacific were engaged during July principally in regrouping, training, rehabilitating, and logistic build-up preparatory to the final assault on the Japanese homeland, Fast Carrier Task Forces and Land-based aircraft embarked upon an intensive program of attacks designed to 'soften-up' the Empire for the final invasion.

In mid-July 1945, land-based Privateers continued to operate from Yontan Airfield Okinawa. Patrol Bombing Squadron 109 returned from rest and repair on Tinian. Patrol Bombing Squadron 109 found,

Yontan Airfield had been considerably altered in the two months of the squadron's absence. Seabees and Army Construction Engineers had exceeded their previous achievements, and the pace of construction had speeded with augmented personnel and equipment. A surfaced strip and adequate parking areas were operationally welcome.

Fleet Air Wing ONE had formalized its operations at the field into a Landplane Detachment, and wing and squadron offices were hosted within Quonset huts. A taxiway ran across the old living area, and CASU(F)-11 camp had been removed to the hills of the old town of CHINA, away from the dust and noise of the strip and work aprons. Some tents had wooden floors, and all had adequate coral against the inevitable mud. Crowding had disappeared with sufficient tents, and mattresses and bedding had appeared. Oil drum showers were available to all personnel, electricity had been installed, and bowsers of drinking water were stationed through the camp. Less frequent alerts and evening air raids made moving pictures possible. Food alone remained mediocre and not greatly improved, but the squadron was generally enthusiastic over the changes.

The enemy had not ceased his attentions to Yontan Airfield, each night generally bringing an alert and bombs falling on the strip on 31 July and 4 August, but personnel paid small attention. CASU, with augmented personnel, managed maintenance smoothly.[45]

Effective 25 July 1945, CASU(F)-11 reported a consistent manning of 44 Officers, and 848 enlisted. In a 25 July 1945, Summary of Operations by Patrol Bombing Squadron (VPB) 118 working from Yontan Airfield they commented, "This squadron wishes to acknowledge the excellent maintenance and supply from CASU(F)-44 (Tinian) and the marvelous cooperation and hard work despite shortages in equipment from CASU(F)-11 (Okinawa)."[46] On this date VPB 118 moved from Yontan Airfield to Yonabaru Airfield.

August 1945

The following was the last CINCPAC Monthly Report of World War II.

CINCPAC *Report of Surrender and Occupation of Japan*, After years of battles and unsuccessful attempts at negotiating a treaty, the United States dropped atomic bombs on Hiroshima (6 August 45) and Nagasaki (9 August 45). A week later, on August 15, Japan announced its intention to surrender. The Japanese foreign affairs minister, Mamoru Shigemitsu, signed the official document on September 2 onboard USS Missouri in Tokyo Bay.[47]

During August CASU(F)-11 continued servicing aircraft at Yontan Airfield. It was also assigned a Carrier Aircraft Pool Detachment located at Awase Airfield which was nine miles across Okinawa near Buckner Bay. Two to three hundred aircraft, or more were under the responsibility of the Officer-in-Charge, CASU(F)-11 Detachment. On 9 August the USS Burleson completed unloading cargo and disembarking all personnel including the 181 men assigned to this Detachment.

The Aircraft Replacement Pool Detachment established at Awase Airfield was similar to many other aircraft concentration points scattered across the Pacific. The primary location for these units was Guam. All of them were established in preparation for the continued war effort against Japan. They were supplied by multiple Escort Carriers making repetitive trips across the Pacific fully loaded with aircraft.

Awase Airfield, Okinawa. Navy Photo.

On 9 August 1945, Lt. Cmdr. Hugh Tollack reported to the Carrier Aircraft Replacement Pool Detachment, Awase Airfield, Okinawa as the Officer-in-Charge. Tollack had just completed his tour on Ulithi as the Commanding Officer of CASU-51. He was already at Awase Airfield when his new Carrier Aircraft Replacement Pool Detachment personnel arrived.

The 145th Naval Construction Battalion was assigned to Yonabaru Airfield Okinawa coincident with the CASU(F)-11 Carrier Aircraft Pool Detachment being assigned to Awase Airfield. In their Battalion War Diary, they reported,

On the night of August 10th, announcement was made of Japan's offer to surrender. All hell broke loose on the island and it rocked like a drunken boat. Most of the 145th personnel not on duty were at the movies, and first impression received from distant ack-ack [anti-aircraft gunfire] tracers was that an air raid was in progress, for although the Japs had lost Okinawa, they had not stopped raiding it. Just before everyone bolted from the scene it was announced that Japan's offer was heard.

For a stifled, silent moment there was utter stillness as everyone was at first stunned at the announcement. They could not comprehend. Then suddenly and spontaneously a cheer, as if from a single mighty throat rent the air.

The sky was a lacework of anti-aircraft fire. Searchlights stabbed their white needles into the clouds in a crazy erratic fashion. Guards cut loose with their Tommies and a hundred others ran for their shacks to grab their carbines and add to the victory din. Even the ships in the harbor threw ack-ack at the stars.

The island went mad. Until an island-wide red alert stilled the celebration. It was a memorable night.[48]

The main contingent of CASU(F)-11, at Yontan Airfield along with its Detachment at Awase Airfield certainly participated in this island-wide moment of outrageous joy!

From 28 July to 15 August 1945, Patrol Bombing Squadron (VPB) 109 was assigned to Yontan Airfield, Okinawa. In its Summary of Operations the following was stated: "Plane maintenance, in the experience of this squadron, by CASUs has been good on the whole, despite shortages of equipment and personnel from which some forward area units have suffered. In this connection, acknowledgement is made of the good work of CASU(F)-44 (Tinian) and CASU(F)-11 (Okinawa)."[49]

September 1945

On 2 September 1945, the peace treaty between Japan and the United States was signed on the quarterdeck of the USS Missouri (BB-63) in Tokyo Bay, Japan.

It had been five months and six days since CASU(F)-11 had landed at Yontan Airfield on Okinawa. On 7 September 1945, CASU(F)-11 moved from Yontan Airfield to Yonabaru Airfield and rejoined ACORN 29. The Army Air Corps now required the longer runways at Yontan Airfield and the Navy aircraft could fly from the shorter runways at Yonabaru Airfield.

Yonabaru Airfield. Navy Photo.

This Airfield was originally established by the Imperial Japanese Army Air Force. On 15 May 1945, the airfield was captured by the U.S. Marines. In June once the fighting had moved further south the 145th Naval Construction Battalion began to improve the airfield for service as a patrol/bomber airstrip. On 15 August 1945, the base with its 6,500-foot runway was ready for use by Navy aircraft. The Airfield was known as the "mud and dust bowl of the Pacific," which was definitely true. The airfield was on the shores of Buckner Bay.

CASU(F)-11 Camp Yonabaru Airfield Okinawa 1945. Navy Photo.

On 9 September 1945 Commander Carrier Division 24 was officially charged with bringing the Pacific portion of the Army, Navy and Marine warriors home from the war. Operation Magic Carpet was the post-World War II operation by the War Shipping Administration to repatriate over 8 million American military personnel from the European, Pacific and Asian theaters. Hundreds of Liberty ships, Victory ships, and troop transports began repatriating Soldiers from Europe in June 1945. Beginning in October 1945, over 370 navy ships were used for repatriation duties in the Pacific with the hope of getting all personnel home by Christmas. Warships, such as aircraft carriers, battleships, hospital ships and large numbers of assault transports were used. The European phase of Operation Magic Carpet concluded in February 1946 while the Pacific phase continued until September 1946.[50,51]

With the conclusion of the war the men of CASU(F)-11 began to ship out. This included those with enough time in the combat zone, orders to other Navy commands, or those whose enlistments were ending. During the next

12 months the size of CASU(F)-11 decreased by 50% from the over 900 officers and enlisted to less than 400 personnel. These were the men who were stationed at both Yonabaru Airfield and the Awase Aircraft Replacement Pool.

On 10 September 1945 the Landplane Unit (TU 50.2.5) resumed operations from the new Fleet Air Wing One base at Yonabaru Airfield. The Privateers that flew for this unit were maintained by CASU(F)-11. When the Landplane Unit departed, their War Diary stated the CASU on which we were based performed as well as the flow of supplies would permit.

On 16 September 1945, the center of Typhoon Ida passed approximately seven miles to the east of Buckner Bay near Yonabaru Airfield. Maximum winds were at 95 miles per hour. On Okinawa most tents were blown away but the metal Quonset huts suffered little damage. There was some loss of life especially at sea where at least one small vessel rolled over and lost the entire crew. On mainland Japan, 2,473 people were killed, 1,283 were missing and 2,452 were injured. Most of the damage occurred in the Hiroshima Prefecture.

On 21 September 1945 Air Base Command Yonabaru took charge of all airfield operations and ACORN 29 was decommissioned. ACORN 54 and CASU-8 were scheduled to move to Awase Airfield, Okinawa, and take charge of the Aircraft Replacement Pool. However, due to the surprisingly sudden end of the war, there was a modification to this plan. ACORN 54 and CASU-8 were diverted to other locations and CASU(F)-11 on Yonabaru Airfield retained responsibility for the Aircraft Replacement Pool which was six miles away at Awase Airfield, Okinawa.

October 1945

By 31 October 1945 Rear Admiral Kendall and his Magic Carpet staff and ships had moved 161,834 very happy men to the West Coast of the United States. Meanwhile a smaller CASU(F)-11 was suffering through a different kind of war. This was a war with Mother Nature and it was horrific in its own way.

In an after action report Commander in Chief, U.S. Pacific Fleet reported on the passage of Typhoon Louise across the island of Okinawa. An extract follows;

On 4 October a typhoon developed just north of Rota in the Marianas Islands. Guam Weather Central called this storm 'Louise' and put out the first weather advisory. Four days later, on the eighth, the storm turned to the right and headed north for Okinawa. All ships and units in the Okinawa area began to be alerted for the storm late in the evening of the 8th. The forecast for Okinawa was for winds of 60 knots, with 90 knot gusts in the early morning of 9 October, and passage of the center along a predicted path 150 miles west of Okinawa.

Louise,' however, failed to follow the forecast, and that evening, as it reached 25 degrees N (directly south of Okinawa) it slowed to six knots and greatly increased in intensity. The actual path taken by Louise was less than 15 miles east of Okinawa's southeast coast. By midday the wind was 60 knots from the east and northeast, with tremendous seas breaking over the ships. The ninth of October found Buckner Bay, a supposedly safe shelter, jammed with ships ranging in size from large Victory ships to small amphibious landing craft. All units, both afloat and ashore, were hurriedly battening down and securing for the storm. However, many small craft were already being torn loose from their anchors, and larger ships were, with difficulty, holding by liberal use of their engines. At 1400 the wind had risen to 80 knots, with gusts of far greater intensity, the rain that drove in horizontally was more salt than fresh, and even the large ships were dragging anchor under the pounding of 30 to 35-foot seas. The bay was now in almost total darkness, and was a scene of utter confusion as ships suddenly loomed in the darkness, collided, or barely escaped colliding by skillful use of engines, and were as quickly separated by the heavy seas. Not all ships were lucky; hundreds were blown ashore, and frequently several were cast on the beach in one general mass of wreckage, while the crews worked desperately to maintain watertight integrity and to fasten a line to anything at hand in order to stop pounding. Many ships had to be abandoned. Sometimes the crews were taken aboard by other ships; more often they made their way ashore, where they spent a miserable night huddled in caves and fields.

Conditions on shore were no better. Twenty hours of torrential rain soaked everything, made quagmires of roads, and ruined virtually all stores. The hurricane winds destroyed from 50% to 95% of all tent camps, and flooded the remainder. Damage to Quonset huts ran from 49% to 99% total destruction. Some of these Quonsets were lifted bodily and moved hundreds of feet; others were torn apart, galvanized iron sheets ripped off, wall boarding shredded, and curved supports torn apart. Driven from their housing, officers and men alike were compelled to take shelter in caves, old tombs, trenches, and ditches in the open fields, and even behind heavy road-building machinery, as the wind swept tents, planks, and sections of galvanized iron through the air.

At the Naval Air Bases some 60 planes of all types were damaged, some of which had been tossed about unmercifully, but most of which were repairable.

The storm center of typhoon 'Louise' passed Buckner Bay at about 1600, from which time until 2000 it raged at peak strength, steady winds of 100 knots and gusts to 120 knots. After 2000 the winds gradually began to subside to winds of 80 and 60 knots throughout the night, and some gusts of higher velocity. A wild, wet, and dangerous night was spent by all hands, afloat or ashore. It was not until 1000 on the 10th that the winds fell to a steady 40 knots, and rains slackened.

On the night of 10-11 October 'Louise' tracked NNE away from Okinawa. This ended typhoon 'Louise', but the damage it left behind on Okinawa was tremendous. Approximately 80% of all housing and buildings were destroyed or made unusable. Very little tentage was salvageable, and little was on hand as a result of previous storms. Food stocks were left for only 10 days. Some of the many ships that washed ashore had been well stocked with meats and vegetables and soon a great deal of bartering was taking place between the ship personnel and the shore-based men. Medical facilities were so destroyed that an immediate request had to be made for a hospital ship to support the shore activities on the island.

Casualties were low, considering the great numbers of people concerned and the extreme violence of the storm. This was largely due to the active and well directed efforts of all hands in assisting one another, particularly in evacuation of grounded and sinking ships. By 18 October, reports had been sorted out and it was found that there were 36 dead and 47 missing, with approximately 100 receiving fairly serious injuries. The casualty list of ships was far greater. A total of 12 ships were sunk, 222 grounded and 32 damaged beyond the ability of ship's companies to repair.

Repair work went on rapidly ashore. As a result of experience in the earlier typhoon in September, extra stocks of food and tentage were to be stored on Okinawa. These were enroute on 9 October, and in less than a week after the storm, supplies were fairly well built up; emergency mess halls and sleeping quarters had been erected for all hands, and 7500 men had been processed for return to the United States.[52]

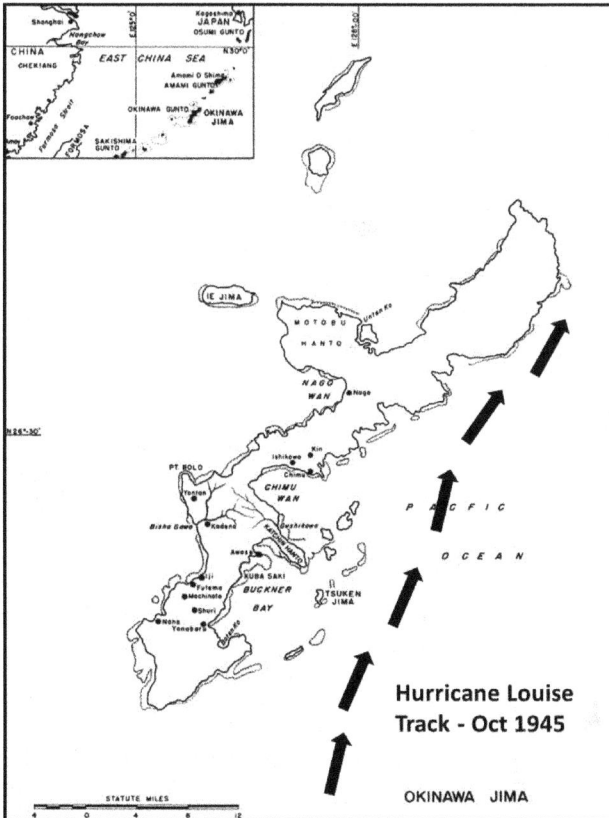

Hurricane Louise, Okinawa Japan. Navy Chart.

On 22 October 1945, the Commander Naval Air Bases, Navy No. 3256 sent the following commendation letter to the Officer-in-Charge, CNAB Aircraft Pool Detachment. This detachment was still part of CASU(F)-11. "From 16 July 1945 until the present date, the officers and men in your charge have been engaged in maintenance and handling of the aircraft pool at Awase Airfield, Okinawa. Your work during this period, part of which has been conducted during combat operations and under constant threat of enemy attack has been of most excellent quality. For the past two months your untiring efforts during intermittent typhoons have been instrumental in keeping the wind damage on pool aircraft at a very minimum. The receipts and issues of aircraft have been accomplished under the most trying conditions, especially with reference to the loading and unloading of aircraft carrier by barge. Your unit has responded admirably to all tasks assigned you, and all of these jobs have been well done. The Commander Naval Air Bases takes pleasure in commending you for this work."[53]

November 1945 – November 1946

On 13 November 1945 Cmdr. Hugh L. Tollack relieved Cmdr. Zimmerman as the Commanding Officer of CASU(F)-11. Cmdr. Zimmerman received the Bronze Star for his service with CASU(F)-11 on Okinawa. "The Bronze Star Medal may be awarded to recognize minor acts of heroism in actual combat or single acts of merit, or meritorious service either in sustained operational activities against an enemy or in direct support of such operations."[54]

CASU(F)-11 continued to maintain aircraft at Yonabaru Airfield and to manage the Aircraft Replacement Pool, Awase Airfield, Okinawa. The manning of CASU(F)-11 at both Yonabaru and Awase Airfields is unknown, however, the number continued to decrease.

On 22 May 1946 Cmdr. Frederic L. Faulkner relieved Cmdr. Tollack as Commanding Officer of CASU(F)-11.

On 11 July 1946, the following message was received by all CASUs. "To establish clear-cut relationships for aircraft maintenance, the Chief of Naval Operations directed the disestablishment of all CASU's and other maintenance units and their replacement by Fleet Aircraft Service Squadrons

(FASRONs) by 1 January. The new FASRONs were to be three kinds according to aircraft types serviced, and were designed to promote higher standards and greater uniformity and efficiency in aircraft maintenance."[55] The CASU replacement date was no later than 1 January 1947.

According to the Naval Aviation News Magazine, "Under present tentative plans air groups will assume responsibility for aircraft, maintenance and operational availability. Fleet Air Support squadrons will replace the former CASU, PATSU, CASD, SOSU, and HEDRON. Air groups and squadrons will become self-supporting for short periods of time. The FASRON will put enlisted men back into squadrons where all work will be under the supervision of squadron C.O.'s."[56]

Since the signing of the peace treaty in September 1945 Naval Aviation had changed significantly. "Since that memorable date the number of Navy personnel on active duty has been cut by half. There was 317,500 men including 49,000 pilots in uniform a year ago, now there are 135,500 men with only 21,500 pilots. Downsizing has been ongoing rapidly, 'better than 11,000 ex-combat planes of assorted designs sit at Clinton, Oklahoma.' Furnaces at Alameda, Jacksonville, San Diego, Corpus Christi, Norfolk, and Miami have been burning part time and full time reducing transport, bomber, and carrier plane fuselages to aluminum ingots."[57]

During the return to peacetime operations the management of aviation personnel, equipment, and aircraft continued. The Navy's Integrated Aeronautics Program (IAP) board, which had previously been concerned with timely production and delivery of aircraft to the fleet, was now in charge of this downsizing effort. The September 1946 issue of the Naval Aviation News Magazine advised, "New aircraft will go directly from the factory to the fleet. After this tour of duty, they will go through overhaul and be sent to the operational training commands and fleet reforming squadrons. The large pool of aircraft left over from war production will be preserved for future use."[58]

As far as the Navy was concerned on 1 September 1946 the Magic Carpet demobilization was complete. The last 127,000 combat troops had been safely delivered home.

"On 27 September 1946 COMFAIRWESTPAC issued the following message.

FROM: COMFAIRWESTPAC
ACTIONS: COMFAIRWINGONE
           COMCASU ELEVEN
           NAB YONOBARU

PLAN DECOMMISSION CASU (F) ELEVEN ABOUT 25 OCT AND ESTABLISH FADRON 122 AT YONABARU COMPLIANCE CNO SERIAL 26P508 OF JULY 11. CONFERENCE REPRESENTATIVES, ACTION ADDEES THIS HEADQUARTERS TUESDAY 15 OCT. ACKNOWLEDGE.

On 9 October 1946, Commander Fleet Air Western Pacific issued temporary additional orders to Cmdr. Faulkner to travel from Yonabaru Airfield to Guam on 14 October 1946. He was to attend a conference to plan the decommissioning of CASU(F)-11 and its re-establishment as Fleet Aircraft Service Squadron 122 (FASRON-122) on or about 25 October 1946. He was further directed to detach from CASU(F)-11 as commanding officer and proceed to FASRON-122 for duty as its Commanding Officer. On 1 November 1946 this was accomplished.

All personnel remaining in CASU(F)-11 were immediately transferred to FASRON-122. Since this action only required a name change the command remained at Yonabaru Airfield, Okinawa. FASRON-122 would linked directly with Fleet Air Wing One which also operated out of Yonabaru Airfield.

Cmdr. Faulkner Report of Orders Executed to FASRON-122.

As of 1 November 1946 CASU(F)-11 ceased to exist as an official command of the United States Navy.

The Naval Aviation News Magazine of October 1946 provided additional details on the transformation of CASUs into FASRONs. The article stated, "Adoption of the Fleet Air Service Squadron (FASRON) under this (IAP) program puts the responsibility for maintenance back into the squadrons."[59] Aircraft maintenance personnel should stay closer to their aircraft, both ashore and afloat. Carriers and airfields would always need some maintenance personnel but not in the amounts prescribed by wartime requirements. As all CASU's, CASU(F)'s, PATSU's, SOSU's, HEDRON's, and CASD's were disestablished its personnel fell into one of three groups, those headed home as having completed their required service time, those headed to aircraft squadrons to perform direct aircraft maintenance and those headed to the new FASRONs.

The service members of CASU(F)-11 were also eligible to receive the Asiatic Pacific Campaign Medal and one engagement star. This was based on their participation in Operation - Okinawa Gunto from 17 March 1945 to 30 June 1945. There were a few men in CASU(F)-11 who served on both Guadalcanal and Okinawa and they earned one ribbon/medal and two engagement stars. The men initially received their ribbons in lieu of the medal because it was too difficult to ship thousands of medals across the combat zone and the recipients were unable to care for them in a combat environment. The men received their medals after the war.[60]

Asiatic-Pacific Campaign Medal and Ribbon.
World War II Victory Medal.

All service members of CASU-11 and CASU(F)-11 were also awarded the World War II Victory Medal. Eligibility was based on those persons who were on active duty at any time between 7 December 1941 and 31 December 1946.[61]

PART Six - Epilogue

This has been the story of a Carrier Aircraft Service Unit, two islands, and two war campaigns wrapped around them. On Guadalcanal CASU-11 had a continuous partnership with its assigned ACORN and then the Naval Air Center. The ACORN and Naval Air Center served as buffers for the men. They were insulated from the bureaucratic noise above and they were able to focus on the work at hand. The men of CASU-11 arrived on Guadalcanal together and left within a month of each other.

CASU(F)-11 endured two disadvantages on Okinawa that impacted its work performance. Within a few weeks its assigned ACORN moved 21 miles away and was occupied building runways as fast as possible on the island. In addition, CASU(F)-11 was put ashore with a small, totally inadequate, advance party of 11 men followed 44 days later by over 800 men. In hindsight, at least 400 men should have gone ashore immediately.

The enemy bombing on Guadalcanal decreased as time passed. During the first couple of months as the Japanese tried to push the Allies off the island, the bombing on Henderson Field was intense. As the Marines moved up the Solomon Island chain taking additional airfields from the Japanese and Allied aircraft increased in number, more and more Japanese aircraft were shot down. This reduced the number of Japanese planes and the number of skilled Japanese pilots in the area.

On the other hand, the Japanese were resolved to remove the Allies from Okinawa, part of Japan's homeland. The Japanese attacking air forces included Kamikaze suicide aircraft by the hundreds. The result was an extremely intense 90 days of almost non-stop bombing with airplanes crashing into any target they could see. This included one Kamikaze attack into the CASU(F)-11 work area at Yontan Airfield. A few weeks later, attacks dropped dramatically.

Guadalcanal was the island of rain, mud and dust, then rain and mud and dust, over and over again. Mosquitos were rampant and this required draining the swamps, oiling the waters, and demanding the men take preventative medication. Most of the men still contracted malaria at least once. On Guadalcanal living conditions generally improved with time.

Okinawa was rain and mud, rain and mud, rain and mud, and fleas. Living conditions were soggy and the housing was poor and temporary. CASU(F)-11 moved twice, once across the airfield, and once from one airfield to another. Repeated typhoons destroyed their efforts at building better housing.

On Guadalcanal repair parts were in thousands of wooden crates spread throughout the coconut plantations. With adequate searching and a hammer the Sailors would eventually find the parts they needed. On Okinawa the parts were on the ships and when the parts were off loaded, they slowly piled up on the beaches. The war was over by the time this situation improved. A typhoon that passed close to Okinawa destroyed many ships, parts warehouses and brought the supply system to a halt until the storm recovery was completed.

Was one location better than the other? No. There were trade-offs. Guadalcanal had mosquitos and malaria, and Okinawa had typhoons and fleas, and both had plenty of bombs falling from the sky.

John Laurence, in his book, *The Cat from Hue,* interviewed a combat veteran and they talked about how long was too long, or too much time in a combat zone. The Vet said, "I figure six months in the field should be enough for any man. But when you have to stay out there ten, eleven months, that ain't no good. You'd never be all to yourself afterwards.[1]" Most of the men of CASU-11 and CASU(F)-11 were in the combat zone for at least 14 months.

Epilogue

Most of the men assigned to CASU(F)-11 came home. Most of them rarely talked about the war. They kept it to themselves. They never forgot about it but they kept it to themselves. Nonstop memories rattled around in their heads, sometimes nightmares, all getting dimmer with the years. A few of them began to tell stories when they reached their eighties and nineties, the last couple of decades of their lives. I sincerely thank those who talked with me and hope I have assembled their stories correctly and coherently. Group photos covering CASU-11's time on Guadalcanal are provided in Appendix A. Additionally, Appendix H provides official Navy comments on CASUs in general and CASU(F)-11 specifically.

Final note: Two of CASU(F)-11's veterans are buried in the Fort Rosecrans National Cemetery, San Diego, California. They are Chief Harry Absolum Hays, Jr. whose diary motivated this book. He is buried in Section CBN, Row 2, Site 207. Cmdr. Frederic Lewis Faulkner is buried in Section CBI, Row 1, Site 380 and his Navy Cross underscores the heroism of every man who served within the ranks of Combat Aircraft Service Unit (Forward) Eleven.

Epilogue

Appendix A - CASU-11 Contributing Members and Group Photos

Chief Hays centered in doorway with cap pulled back.

Harry Absolum Hays, Jr.

Born - 19 September 1910, Peru, Indiana.
Enlisted - 28 May 1928.

Duty Stations - CASU-11, Naval Air Station Oakland, California, Air Transport Squadron Four.

Separated - 6 October 1945, Separation Center Shoemaker, California.

Death - 7 Apr 2009, interred Ft. Rosecrans National Cemetery San Diego, California. Section CBN, Row 2, Site 207.

Juan Leal, Jr.

Born - 23 May 1924, La Paloma, Texas
Enlisted - 29 October 1942
Duty Station - CASU-11
Death - 16 October 1943, Guadalcanal. After his accidental death at CASU-11 Guadalcanal SC3c Leal was buried in the Army, Navy, Marine Cemetery Guadalcanal, Grave 1, Row 73. According to the Brownsville Herald, Brownsville, Texas his remains were later moved to La Paloma Cemetery, La Paloma, Texas. A veteran's gravestone was requested by the family.

Robert Henry Little

Born - 15 March 1920, Whiteford, Maryland
Enlisted - 13 January 1942
Duty Stations - CASU-11, Naval Air Station Corpus Christi, Texas
Retired - 15 June 1962 as ADJ1 (Aviation Machinist Mate (Jets) First Class)
Death - 2 March 2003, Westminster, Maryland

John Joseph McAteer

Born - 24 September 1920, Philadelphia, Pennsylvania
Date of entry - 16 August 1942
Duty Stations - CASU-11, Ammunition Depot, Hingham, Massachusetts, USS Thetis Bay (CVE-90)
Separated - 17 August 1946
Death - 19 January 1996, Camden, New Jersey

Patrick Gregory O'Flynn

Born - 4 January 1924, Laurel, Mississippi
Enlistment - 27 October 1942
Duty Stations - CASU-11, Naval Station Jacksonville, Florida
Retired - 4 August 1963 as ENC (Chief Engineman)
Death - 8 July 2017, Tampa, Florida

Henry Franklin Parker

Born - 18 February 1922, Circleville, Kansas
Enlistment - 11 September 1942
Duty Stations - CASU-11, CASU-6 at Naval Air Station Alameda, California
Separated - 23 October 1945
Death - 24 January 2010, Discovery Bay, California

Durward West

Born - 19 July 1923, Lexington, Illinois
Enlistment - 28 August 1941
Duty Stations - CASU-11, CASU-12, CASU-50, Naval Air Station
Corpus Christi, Texas
Separated - 6 December 1946
Death - 18 June 1991, Palatine, Illinois

David Anderson Wilson

Born - 11 April 1921, Ludlow, Mississippi
Enlistment - 7 February 1942
Duty Stations - CASU-11, CASU-8, CASU-6, Naval Air Station, Alameda, California
Separated - 13 December 1945
Death - 22 August 1999 Hackensack, New Jersey

CASU 11 Group Photos Guadalcanal

Fifth Division, Supply. From John McAteer.

In late May 1943 Lt. McAteer joined CASU-11 after it had arrived on Guadalcanal. McAteer is kneeling in the middle of the front row (fourth from the right) in a long sleeved shirt rolled to mid-forearm. He wrote the following on the back of this picture of the Fifth Division, Supply;

"The two men behind me are S/K Keimig SK2c, Weaver S/K2c; first line on men on far right - Pantry S/K1c. First line of men on far left Leitle S/K1c, Raine S/K1c - all work for me in Disbursing office. White & Smith. Chiefs. (In pith helmets.) Lt. Dickinson, Supply Officer (Long sleeved shirt rolled up to mid bicep.). July 1943. Picture is of the Eight Fifth Division."

"S/K" stands for storekeeper. Interestingly the Supply Division was labeled the Fifth Division. Standard Navy Organization for CASUs lists the supply group as the Sixth Division. This reveals that Commanding Officers would reorganized their commands to meet the circumstances.

CASU-11 SBD Engineering Gang. From Robert Little.

Robert Little is in the four man group centered in picture. This is the fourth row and he is second from the right. Or you can go to the bare chested man at the top center of picture. His right arm drops down behind my father's head.

CASU 11 Motor Pool. From Patrick O'Flynn.

Patrick O'Flynn is the highest man in the right half of this picture.

CASU 11 Metalsmiths. From Harry Hays.

David Wilson is the second man from the right, kneeling in the second row.

Frank Parker is the sixth man from the left, standing bare chested in the third row (the first of the standing rows).

Harry Hays is on the right, fourth row up, very small, no shirt, hairy chest. Harry is the right hand most man in a row of three with the man to his immediate right wearing a white T-shirt.

CASU-11 Radio and Electronics Gang. From Durward West.

Durward West is in the second row, second from the left.

Appendix B - Chief Metalsmith Harry A. Hays Joins CASU-9

This appendix details the last few weeks that Chief Metalsmith Harry Absolum Hays spent in the Southwest Pacific. It ends with his trip home. Hays was now stationed at CASU-9 located at Banika Airfield on Mbanika Island in the Russell Islands. The Russell Islands are located 50 nautical miles to the northwest of Guadalcanal.

Banika Airfield Russell Island 1944. Navy Photo.

Harry's diary continues.

May 1944

Mon. 1ˢᵗ Quiet. Got my orders. Leave tomorrow. Tues. 2ⁿᵈ Quiet. Flew up on B-25. Looks bad. Mud, no drinking water, poor chow, etc. Sure will be glad when my 18 months are up. Will know about working conditions tomorrow. I think we work every day, and until 4:00. And no beer either. Wed.3ʳᵈ Quiet. Yep, everyday, from 7:30 to 4:30. Boy, I really got stuck - sure wish I was back in good old C.A.S.U. #11. This is really awful! Thurs. 4ᵗʰ Quiet. Lots of work to do. Sure hope I can handle my end of it. Fri. 5ᵗʰ Quiet. I'm in charge of line crew. Hoping for the best. Sat. 6ᵗʰ Quiet. All O.K. So far. Sun. 7ᵗʰ Quiet. Went to Mass. Working from 7:30 till 4:30. Mon. 8ᵗʰ Quiet. Working from 7:30 till 5:00. Tues. 9ᵗʰ Quiet. Pretty well caught up. Wed. 10ᵗʰ Working from 7:00 till 5:00 with 1 hour for lunch. Have 12 to 4 tonight. Boy, this is rough duty! Thurs. 11ᵗʰ Quiet. Well, I made it, but I'm sure pooped! Fri. 12ᵗʰ Quiet. Same hours. Sat. 13ᵗʰ Quiet. Sun. 14ᵗʰ Quiet. Went to Mass. Finished planes today now we're starting to build a quanset hut-whew! Have 4 to 8 tomorrow A.M. Mon. 15ᵗʰ Quiet. My name is supposed to have gone in today for relief. Boy, I hope it did! Tues. 16ᵗʰ Quiet. Wed. 17ᵗʰ Quiet. First day off. Thurs. 18ᵗʰ Quiet. Fri. 19ᵗʰ Quiet. Have 8 to 12. Sat. 20ᵗʰ Quiet. Sun. 21ˢᵗ Quiet. Went to Mass. Have day off. Mon. 22ⁿᵈ Quiet. Tues. 23ʳᵈ Quiet. Have 12 to 4. Just learned that my name won't go in for relief until next month. Damn it! Wed. 24ᵗʰ Quiet. Thurs. 25ᵗʰ Quiet. Fri. 26ᵗʰ Quiet. Sat. 27ᵗʰ Quiet. Sun. 28ᵗʰ Quiet. Went to Mass. Mon. 29ᵗʰ Quiet. Tues. 30ᵗʰ Quiet. Wed. 31ˢᵗ Quiet. From the diary of Harry Hays, May 1944.

Quonset Huts (Normally all metal construction.). Navy Photo.

June 1944

Thurs. 1ˢᵗ Quiet. Pay day. Have $510.00 on books. Fri. 2ⁿᵈ Quiet. Have 12 to 4. Sat. 3ʳᵈ Quiet. I get ½ set of flight skins. Sun. 4ᵗʰ Quiet. Went to Mass. Mon. 5ᵗʰ Quiet. Tues. 6ᵗʰ Quiet. Wed. 7ᵗʰ Quiet. Thurs. 8ᵗʰ Quiet. Have 4 to 8. Fri. 9ᵗʰ Quiet. Went to the Canal on TBF. Had visit with old gang. Sat. 10ᵗʰ Quiet. Returned to C.A.S.U. #9 after all night in Canal. Swell trip! Sun. 11ᵗʰ Went to Mass. Quiet. Mon. 12ᵗʰ Quiet. Tues. 13ᵗʰ Quiet. Supposed to be a Jap task force near. All B-25s and P- 38s are grounded, but all loaded and ready to go. Wed. 14ᵗʰ Quiet. Have 8 to 12. Thurs. 15ᵗʰ Quiet. No dope on Jap ships. Man off post last night. Have to go to mast. Fri. 16ᵗʰ Quiet. Sat. 17ᵗʰ Quiet. Sun. 18ᵗʰ Quiet. Went to Communion. Mon. 19ᵗʰ Quiet. Tues. 20ᵗʰ Quiet. Wed. 21ˢᵗ Quiet. Thurs. 22ⁿᵈ Quiet. Fri. 23ʳᵈ Quiet. From the diary of Harry Hays, June 1944.*

Hays' position required some flight time, probably to test repaired aircraft. Hence his pay was increased by adding one half of "flight skins" which was the additional pay that all pilots received. Canal is short for Guadalcanal, and a TBF is a Bomber/Fighter aircraft with three seats.

When a mild infraction of the rules occurred, the accused party was sent to "Captain's Mast" for adjudication. Minor infractions included late to work or not being at your assigned post. The Captain gathers the accused, the accuser, witnesses, and those who can speak to the accused man's performance of his duties. Harry, as the man's Chief, probably went to the Mast as a person who could testify about the man's day-to-day work performance.

Sat. 24ᵗʰ Quiet. Have 8 to 12. Sun. 25ᵗʰ Quiet. Went to Mass. Sure was excited today. Supposed to be an exchange of men between C.A.S.U. #8 and #9, all men with 14 months out to go home. Also heard that another Metalsmith Chief was coming here. Now I hear that it has all been cancelled. Damn it! Mon. 26ᵗʰ Quiet. Tues. 27ᵗʰ Quiet. Wed. 28ᵗʰ Quiet. Oh Boy! Heard today we're going back [Home.] within 4 days. Whoopee!! Have $589.00 on books. Am drawing it all on a check. Thurs. 29ᵗʰ Quiet. Could only draw $50.00, leaves balance of $539.00 on books. Fri. 30ᵗʰ Am all packed. Just heard we won't go aboard until tomorrow. I have the 4 to 8 tonight, too. From the diary of Harry Hays, June 1944.

July 1944

Sat. 1st Quiet. Missed the ship. Now we have to wait for another. Damn it, anyway! Sun. 2nd Quiet. Went to Mass. Mon. 3rd Quiet. Tues. 4th Quiet. Just heard we may board a small carrier today or tomorrow. Should be a better and faster trip. Wed. 5th By Golly, we had an alert last night. Probably was a plane with no I.F.F. No one got up as all of the fox-holes have been closed, filled in. Thurs. 6th Quiet. Fri. 7th Quiet. Sat. 8th Quiet. Boy I wish we'd get out of here. Sun. 9th Quiet. Went to Mass. Mon. 10th Quiet. Had 4 to 8. Tues. 11th Quiet. Wed. 12th Quiet. Still standing by, and it's sure getting monotonous. Thurs. 13th Quiet. Fri. 14th Quiet. Sat. 15th Quiet. Have 4 to 8. Sun. 16th Quiet. Went to Communion. From the diary of Harry Hays, July 1944.

The STAG-1 was also known as the TDR-1 "Torpedo Drone." It was the first drone equipped with a first-generation TV camera and radio remote controls. It could be flown remotely by pilots aboard chase aircraft. They could strike the enemy at will without risking American lives. When they were finally placed into combat conditions, 46 drones were launched with nearly 50% of the strikes judged a success, and no American lives were lost.

TDR – 1 Torpedo Drones awaiting shipment overseas. Navy Photo.

Mon. 17th Quiet. Worked all day at Stag #1. Hooray! At about 4:30 P.M. we got word to be ready to board Navy Transport in 2 hours. This is what we have been waiting for! Finally all aboard at around 12:00 P.M. C.A.S.U. #19 also coming aboard. Swell stateroom, 6 man, but pretty hot. Lockers too! Lots better than coming out. U.S.S. General Harry Taylor. Tues. 18th Shoved off at about 9:00 A.M. It is now 11:30 and we are off of the Canal. Going to pickup more troops. It is now 5:30 P.M. And we just shoved off for HOME AND OLIVE! From the diary of Harry Hays, July 1944.

U.S.S. General Harry Taylor (AP-145). Navy Photo.

Wed. 19th Didn't get away until 6:30 A.M today. But we are really under way now. Sure hope we have no more stops. Compared to our trip over this is a luxury liner. Good chow, 6 man compartment, good bunks, fresh water, long showers. Compartment hot, but it will get better. Thurs. 20th Heard today we are trying to make it in 11 days. Hope so! Our escort of 1 destroyer left us today at noon. We are now on our own. Fri. 21st All secure. Sat. 22nd. Quiet. Sat. 22nd Same day again as we crossed the 180th meridian. Sun. 23rd Quiet. Went to Mass. Mon. 24th Quiet. Tues. 25th Quiet. Had a little scare last night. We stopped for about 20 minutes - something wrong with the electricity. Could have been bad! Wed. 26th Quiet. Crossed equator at about 4:00 P.M. Today. Thurs. 27th Quiet. Fri. 28th Quiet. I have dysentery again. Sat. 29th Quiet. Took medicine for dysentery. Sun. 30th Quiet. Have bad cold. Doctor told me to stay in bed so no Mass. Mon. 31st Quiet. Am constipated now so had to take laxative. I'm having a hell of a time! From the diary of Harry Hays, July 1944.

August 1944

Tues. 1st Quiet. Feel much better today. I hear we will be in Thurs. A.M. Early. Sure hope so! Wed. 2nd Quiet Tomorrow is the day! THURS. 3RD OF AUGUST. It's now about 6:00 A.M and we can see the lighthouse. Seagulls, too! It sure looks good!!! Last entry. From the diary of Harry Hays, August 1944.

Late on the 3rd of August 1944, the USS General Harry Taylor arrived at the Matson Navigation Company piers on the San Francisco waterfront. Olive, the love of Harry's life, met him in San Francisco. Harry's sister, Mary, said, "They spent a couple of weeks in the city before taking the train to San Diego." When Harry and Olive got off the train, Mary said, "He looked so handsome in his uniform."

A few days later Harry and Olive returned to the bay area as Harry had orders to Air Transportation Squadron Four located at NAS Alameda, California. On 12 October 1945 he ended his Naval Service.

Harry's Poem

I'm not tall,
In fact I'm kinda short.
I'm not mean,
I'm quite a peaceful sort.
I hate to admit
I've never had a fight
Mussolini can't be right.

I'm not strong,
In fact I'm kinda weak.
I'm not brave,
In fact I'm kinda meek.
I'm in between;
I'm neither man nor mouse
But I'm not afraid to say
Schickelgruber Hitler is a louse.

I'm not well,
In fact I'm kinda sick,
But those two guys alone-would be
a cinch to lick.
So they called in a partner-to take
his turn at bat-
A mug called Hirohito,
A squint-eyed little rat.

Now I'm not the only sailor,
So please don't get me wrong.
Cause there are lots of other sailors
-Plenty big and strong.
Who'll back me up in what I say.
And when it comes to pass-
You can bet your
bottom dollar,
We'll knock them on their ear.

H.A. HAYS, Jr. AM2c,
C.A.S.U. 11 - Naval Air Station

Chief Metalsmith Harry A Hays Joins CASU-9

Appendix C - Commanding Officers of CASU-11

The following in chronological order were the Commanding Officers of CASU-11. All were qualified aviators who flew a variety of aircraft in support of war-related missions. One received the Navy Cross and two received the Bronze Star for their bravery and heroism.

They averaged 41 years of age, most had joined the service well before the war commenced and stayed in the service for years after. Except as noted all the information provided came directly from the military service records of each of these officers.

Commander Isaac Schlossbach (SN 8973) in command - 22 January 1943 to 17 April 1944. From NAS San Diego, California, he traveled with CASU-11 to Henderson Field, Guadalcanal.[1]

Commander Carleton Maxwell Pike (SN 26055) in command - 17 April 1944 to 5 July 1944. CASU-11, Henderson Field, Guadalcanal.[2]

Lt. Commander Reinhart E. Vogt (SN 62360) in command - 5 July 1944 to 22 July 1944. CASU-11, Henderson Field, Guadalcanal.[3]

Lt. Commander Philip Allen Jr (SN 37335) in command - 22 July 1944 to 3 August 1944. CASU-11, Henderson Field, Guadalcanal.[4]

Lt. Raymond B. Torian (SN 118179) in command - late August 1944 to 23 December 1944. CASU-11 was at ACORN Assembly and Training

Detachment, Port Hueneme, California. Assigned as, "Acting Commanding Officer."[5]

Commanding Officer unknown in command - 7 December 1944 to 23 December 1944. Sadly, committed suicide during his tenure as commanding officer.[6]

Commander E. F. Zimmerman (SN 72029) in command - 23 December 1944 to 13 November 1945. CASU(F)-11, Naval Air Facility Thermal, California, traveled to Yontan/Yonabaru Airfields, Okinawa, Japan.[7]

Commander Hugh L. Tollack (SN 170496) in command - 13 November 1945 to 4 March 1946. CASU(F)-11 at Yonabaru Airfield, Okinawa, Japan.[8]

Commander Charles E. Hays (SN 149092) in command - 4 March 1946 to 22 May 1946.   CASU(F)-11 at Yonabaru Airfield, Okinawa, Japan.[9]

Commander Frederic L. Faulkner (SN 79572) in command - 22 May 1946 to 1 November 1946. CASU(F)-11 Yonabaru Airfield, Okinawa, Japan.[10]

Cmdr. Isaac (Ike) Schlossbach

Midshipman Isaac Schlossbach. Navy Photo.

On 7 December 1941, Pearl Harbor Day, the day the first bombs of WWII fell on Hawaii, Isaac (Ike) Schlossbach was 50 years, 6 months and 17 days old. On 22 January 1943, Cmdr. Schlossbach commissioned CASU-11 and became its first Commanding Officer. This WWI veteran was now 51 years old and was again headed to war.

According to his military service record, Schlossbach was born on 20 August 1891 in Bradley Beach, New Jersey. He was raised in Neptune Township, New Jersey, and he attended Neptune High School. According to Wikipedia, in 1911, Schlossbach was "the first Jewish midshipman at the U.S. Naval Academy in Annapolis, Maryland, winning letters in football and wrestling. He graduated from the Naval Academy in 1915 and volunteered for the first submarine school. During WWI he commanded submarines in the Mediterranean."[11] As recorded in his military service record, he survived a submarine collision.

In 1921, after the end of World War I, Schlossbach joined the aviation branch of the U.S. Navy. He was first sent to lighter-than-air flying school (dirigibles). In 1922 he learned to fly fixed-wing aircraft. By 1925 Schlossbach was leading an aero squadron. He commanded the first squadron to serve on the first aircraft carrier, the USS Langley. Later Schlossbach had trouble with his left eye, and the Navy assigned him to the Naval Academy. He taught engineering, aviation and coached the football team. In 1930 at the age of 38 Schlossbach had his left eye surgically removed and he was forced to retire on a medical discharge.

Ike aboard the schooner A.W. Greely, 1937.

As a civilian Schlossbach "went on twelve polar expeditions, three to the Arctic and nine to the Antarctic. He was on the Wilkins Trans-Arctic Expedition in 1931 and served as navigator on the USS Nautilus during the first attempt to take a submarine to the Pole under the icepack. He commanded Admiral Richard Byrd's ship the *Bear of Oakland* and was a pilot on Byrd's Second Antarctic Expedition (1933–35). He was second in command on the MacGregor Arctic Expedition (1937-38) where he accomplished a number of polar aviation firsts. In 1939, he again accompanied Byrd to the Antarctic on the United States Antarctic Service

Expedition. On 10 October 1941, just before the U.S. entry into World War II, Schlossbach headed a small radio/meteorological team that founded the airport at Fort Chimo (Crystal I) in northern Labrador."[12]

On 5 June 1939, Schlossbach was recalled to active duty. However due to dental issues he failed his physical, and only after getting his dentures replaced did he gain entrance into active service. On 30 December 1942 he arrived at Naval Air Station San Diego with orders "to be the Commanding Officer of CASU-11 when unit commissioned." CASU-11 was commissioned into operational status on 22 January 1943 with Schlossbach as its first Commanding Officer.

Schlossbach took CASU-11 to Guadalcanal where he served simultaneously as Commanding Officer of both CASU-11 and Henderson Field. "After WWII, Schlossbach was second in command on the 1946-1948 Ronne Antarctic Research Expedition the last privately funded Antarctic expedition. He commanded the 1,200-ton, diesel-powered, wooden tug Port of Beaumont which was frozen in Back Bay through the winter. He also accompanied Finn Ronne to a cape in the Weddell Sea. A cape and a mountain were named after Schlossbach. In 1955 Schlossbach accompanied an Australian research expedition to Ellsworth Station in Antarctica, for which he received a letter of commendation from the Australian government. In 1956, Admiral Byrd selected Ike as his personal representative in Operation Deep Freeze. In 1961 Schlossbach accompanied Byrd on several other occasions and made his last trip to the Antarctic when he was 70 years old as a consultant to the U.S. Navy. Schlossbach was awarded three Congressional Medals for his contributions to Antarctic exploration."[13] He never married, and his August 1984 obituary stated he was an "adventurer who made 12 polar expeditions, fought in two world wars and survived four airplane crashes before passing at the age of 93."[14] This was quite an impressive record for the first Commanding Officer of CASU-11.

Cmdr. Carleton Maxwell Pike

Cmdr. Carleton Maxwell Pike. Navy Photo.

Pike, a native of Lubec, Massachusetts was born on 9 October 1894 and died on 25 April 1967. "He served in both WWI and WWII. In 1916 he went directly from Maine's Bowdoin College to France as an ambulance driver. In April 1917 he returned, graduated from Bowdoin, and then joined the Naval Reserve aviation program. Two years later he reached the rank of Lieutenant Senior Grade. He was among the first 200 men qualified as aviators"[15] within the U.S. Navy.

Between the wars he had a career in business. During WWII he rejoined the Navy and became the Commanding Officer, Seaplane Base Navy #131, Ile Nou, New Caledonia. This was followed by Assistant Operations Officer at Fleet Air Wing One Staff Munda, New Georgia. Later he served as the Commanding Officer of CASU-11, Henderson Field, Guadalcanal. He served a total of 19 months in the South Pacific. Pike's final tour in the U.S. Navy was as the executive officer at Whidbey Island Naval Air Station in Washington State. On 1 October 1945, he retired at the rank of Commander.

Cmdr. Reinhart Edmund Vogt

Cmdr. Reinhart Edmund Vogt. Family Photo.

He was born in Massillon, Ohio, on 2 August 1908 and died in Sarasota, Florida at the age of 73. On 9 August 1921, he enlisted in the Navy and he was released from active duty on 1 September 1968. Before leaving the Navy Reinhart attained the rank of Captain. During his tour with CASU-12, operating near Augusta Bay, Bougainville, he earned a Bronze Star. His heroic action was not explained in his service record. On 5 July 1944 he became Commanding Officer of CASU-11, and he had the second shortest tour in that position. It lasted only two and a half weeks.

Lt. Cmdr. Philip Allen Jr

Lt. Cmdr. Philip Allen Jr. Navy Photo.

He was born in November 1895 in Warwick, Rhode Island, and on 19 January 1963 at the age of 68 he died in Media, Pennsylvania. Upon graduation from high school Allen studied at Yale University. On 21 January 1918 before his graduation he joined the Navy and qualified as a Pilot. He served in both World War I and II. After WWII he returned to Yale and completed his Bachelor of Arts degree. His stint as Commanding Officer of CASU-11 lasted 12 days and it was the shortest of all commanders. During World War II he served as Commanding Officer of CASU-41, CASU-11, and CASU-42. On 1 August 1957 he retired from active duty.

Lt. Raymond B. Torian

Information on Lt. Torian is very limited. He was born 22 February 1920 in Los Angeles, California and died 4 January 1979. On 11 December 1941, he joined the Navy and soon thereafter qualified as a Navy Pilot. He was assigned as "Acting Commanding Officer" of CASU-11 from August to December 1944, and at age 23 he was the youngest man ever in that position. I became aware of Lt. Torian from the service record of Torpedoman's Mate Felix Durand. Lt. Torian identified himself as the "Acting Commanding Officer" for CASU-11 and signed a letter on 25 September 1944 acknowledging the arrival of TM Durand at CASU-11.

Unknown Officer

The CASU-14 War Diary dated 8 November 1945 reported the following information. "Shortly after arriving at Thermal, the Commanding Officer of CASU-11 committed suicide which also added to the general feeling of dissatisfaction and unrest."[16] CASU-11 arrived at Naval Air Facility Thermal, California on 7 December 1944.

On 14 December 1944, Cmdr. Zimmerman, the next commanding officer for CASU-11, was detached from the staff of Commander, Fleet Air West Coast. His orders to CASU-11 must have surprised him. The previous June he had checked in for at least a year's duty at Fleet Air West Coast. Surely he was advised that the suicide was the cause for his unexpected orders.

Cmdr. Zimmerman assumed command of CASU-11 on 23 December 1944. Consequently, the suicide of the unknown officer must have occurred between 7 and 23 December 1944.

Cmdr. Eugene F. Zimmerman

Cmdr. Eugene F. Zimmerman. Navy Photo.

Cmdr. Zimmerman was born on 17 November 1903 in Pittsburgh, Kansas. His obituary notes: "In 1925, he graduated from Washington University in St. Louis with a degree in electrical engineering. After working for General Motors, in 1930 Zimmerman joined the aviation department of the Shell Oil Co. in St. Louis. In 1932 he joined the U.S. Naval Reserve, and in 1945 he was assigned to temporary active duty as a member of the Naval Reserve Policy Board in Washington, D.C. He participated in the occupation of Okinawa and was awarded the Bronze Star. He returned to the United States in 1946 after assisting an airline company survey routes to the Far East in China, Korea and the Philippines. Zimmerman was a commander of Naval Air Reserve Fleet Air Service Squadron 164 stationed at Floyd Bennett Field until retiring in 1962. Zimmerman was married to Isabelle Lois Zimmerman. He was member of the Staten Island Power Squadron and the Quiet Birdsmen, a national organization of pilots. He was an elder of the First Presbyterian Church, Stapleton, New York. Zimmerman died on 7 December 1967."[17]

On 23 December 1944, Zimmerman took command of CASU-11 at Naval Air Facility, Thermal, California. He traveled with CASU-11 to Yontan/Yonabaru Airfields, Okinawa, Japan. At his death he was only 64 years old.

Cmdr. Hugh L. Tollack

Cmdr. Hugh L. Tollack. Navy Photo.

Cmdr. Tollack was born on 7 May 1903 in Black River Falls, Wisconsin. In 1927 he graduated from the University of Wisconsin with a Bachelor of Arts. In 1939 he completed his law degree at the Minneapolis College of Law. He graduated Magna Cum Laude. Tollack was 39 years old when he entered the Navy. After Flight School and prior to receiving orders to become the officer-in-charge of the Aircraft Replacement Pool, Awase Airfield, Okinawa, he was the Commanding Officer of CASU-51 and CASU-45. After four months in the Aircraft Replacement Pool leadership position, he become the eighth Commanding Officer of CASU-11. On 15 July 1954 Tollack was discharged from the Navy. He passed away on 2 January 1987 at the age of 84.

Cmdr. Charles Elder Hays

Cmdr. Charles Elder Hays. Navy Photo.

On 3 July 1902, Cmdr. Hays was born in Trenton, Tennessee. Prior to WW II, Hays flew for the Royal Canadian Air Force. He spent 18 months with them, and on 3 June 1942 he was accepted into the U.S. Navy as a pilot with the rank of Lieutenant. His tours included Operations Officer of CASU-22 at Charleston, Rhode Island, Commanding Officer of CASU-27 at Naval Base, Quonset Point, Rhode Island, and Executive Officer at Naval Air Station, Astoria, Oregon. He then went to Okinawa to assume command of CASU(F)-11. Hays' last tour in the Navy was on the Staff of the Naval Air Transport Service (NATS). His request to remain on active duty was denied and on 14 August 1949 he was released. On 1 October 1955, he retired from the Naval Reserves. He was married to Carol Raney Hays and they had four children. On 27 November 1971 Cmdr. Hays died at the age of 69.

Below is a picture of Hays while he was serving as Commanding Officer of CASU(F)-11. His name was misspelled in the description of the picture; this mistake occurred regularly in his service record.

These officers direct units making up Yonabaru Naval Air Base. From left to right, Comdr. G. W. Sloan, NAB; Comdr. Charles Hayes, CASU F 11; Lt. Comdr. G. E. Hoffman, VPB 108; Lt. Comdr. J. D. Seal, VPB 128.

Provided by Donn at www.rememberingokinawa.com.[18]

Cmdr. Frederick Lewis Faulkner

Cmdr. Frederick Lewis Faulkner. Navy Photo.

Cmdr. Faulkner was born on 17 July 1914 in Morristown, New Jersey. He graduated from Rutgers University with a liberal arts degree. He was a member of the Delta Phi Fraternity and was a springboard diver on the varsity swim team. On 28 December 1937, Faulkner joined the Navy, successfully completed flight school and then joined the fleet as a qualified pilot.

During May 1942 Faulkner was part of the Scouting Squadron Five (VS-5) on board the USS Yorktown, and he participated in the Battle of the Coral Sea. His heroic actions on 4 and 8 May against the Japanese fleet earned him a Navy Cross. Admiral Halsey personally presented the Navy Cross to Faulkner. This award is second only to the Congressional Medal of Honor. His service was set forth in this citation:

The President of the United States takes pleasure in Presenting the NAVY CROSS to

LIEUTENANT JUNIOR GRADE FREDERIC L. FAULKNER UNITED STATES NAVAL RESERVE

For service as set forth in the following

CITATION:

For extraordinary heroism and courageous perseverance as pilot of an airplane of a Scouting Squadron Five (VS-5), attached to U.S.S. Yorktown (CV-5), in two dive bombing attacks, the first on enemy Japanese forces in Tulagi Harbor on May 4, and the second on an enemy aircraft carrier in the Coral Sea on May 8, 1942. Pressing home these attacks in the face of tremendous anti-aircraft fire and, on May 8, also harassed by heavy aircraft opposition, Lieutenant (jg) Faulkner assisted greatly in the sinking or severe damaging of the carrier and eight other enemy vessels. His conscientious devotion to duty and gallant self-command against formidable odds contributed materially to the success of our forces in the Battle of the Coral Sea.[19]

For the President,
JAMES FORRESTAL
Acting Secretary of the Navy

On 9 May 1942, however, Faulkner was reminded that heroes are also human, and subject to human errors. Around 0900 Faulkner, while flying a surveillance mission, spotted what he thought was an enemy carrier force below him on his left side. The enemy ships were 175 miles from the USS Yorktown and headed east at a speed of 25 knots. Faulkner attempted to make a radio report, however, his radio cut out in the midst of reporting this sighting, leading listeners in the Yorktown to speculate whether he was shot down. His wingman, Ensign Laurence G. Traynor, completed the message. Then the two airmen headed back to the ship expecting to land in about thirty minutes, but no ship was found. After flying for almost two hours they sighted the broad oil slick that trailed 50 miles in the wake of the damaged Yorktown. Faulkner and Traynor finally found their carrier around 1130.

Faulkner quickly briefed the admiral and his staff. Admiral Fletcher requested an immediate carrier strike which after a long search sighted nothing but a "mesa type reef," where breaking surf resembled a ship's wake. Fletcher "asked Faulkner if he could have been mistaken and, upon retrospection, Faulkner decided that he had.[20]

Faulkner was the last commanding officer of CASU(F)-11. On 1 November 1946 CASU(F)-11 was disestablished and reconstituted as Fleet Aircraft Service Squadron (FASRON) 122. Subsequently Faulkner became the first Commanding Officer of FASRON-122 and he held the position until he was relieved on 19 April 1947.

He was later promoted to Captain in the Navy Reserves and worked as Assistant Director in the Navy Bureau of Aeronautics, Weapons Support Systems, until his final retirement. On 28 June 2007 Faulkner died; he was 93 years old. He is buried at Fort Rosecrans National Cemetery, Section CBI, Row 1, in Site 380, San Diego, California.

Appendix D - Awards and Commendations

Commendation Letters and Messages

The following commendation messages and letters can be found in many of the CASU-11 service members' records but not all. All who served with CASU-11 during its Guadalcanal assignment, March 1943 through August 1945, should have had the following citations included in their personal service records.

Most of the laudatory comments below were provided to all supporting forces participating in the Solomons Campaign. Once these were received by higher commands, they were passed down through the various military staffs until they reached each Sailor and Marine.

Similar commendation documents for the Okinawa period have also been included in the text. And, likewise, all who served with CASU(F)-11 during their Okinawa combat mission, January 1945 through October 1945, should have the Okinawa letters attached to their service records.

1. **Commendation Letter**

25 July 1943

From: Commander Naval Bases, SOLOMON ISLANDS
To: All Activities under this command.

Subject: Commander Air, SOLOMONS, Dispatch 240930 of July, 1943.

Subject dispatch is quoted herewith:

"YOUR UNFAILING AND UNSTINTED COOPERATION AT ALL TIMES AND TO ALL UNITS OF MY COMMAND HAS BEEN OF UTMOST HELP IN THE JOB OF KILLING JAPS WARMEST THANKS AND BEST WISHES FOR BETTER DAYS TO COME MITSCHER SENDS XX"

It is with pleasure that Commander Naval Bases, SOLOMONS, quotes the above dispatch, believing that the commendatory remarks contained therein apply to all officers and men under this command.

/s/ W. M. QUIGLEY

This was followed a few days later by another message from Admiral Mitscher:

"MITSCHER AND HARRIS SEND TO THE BEST AIR FORCE WE KNOW AND THE AIR FORCE BEST KNOWN TO THE JAPS X WE ARE SAD AT RELINQUISHING COMMAND OF THE TOUGHEST AND BEST BUNCH OF JAP KILLERS ON RECORD X BEST OF LUCK THE BEST OF LANDINGS AND THE BEST OF HUNTING X GOOD BYE GOOD LUCK AND GOD BLESS YOU.

2. **Commendation Message**

COMAIRSOLS   SEPTEMBER 7, 1943

CONGRATULATIONS FOR YOUR EXCELLENT BOMBING OF MUNDA

3. **Commendation Message**

FROM: COMAIRSOLS   SEPTEMBER 10, 1943

TO: STRIKE COMMAND

FOR DESTRUCTION OF NIP GUNS ON KOLOABANGARA MORNING OF THE 9TH COMAIRSOLS SENDS WELL DONE FOR YOUR EXCELLENT BOMBING

4. Commendation Letter

HEADQUARTERS COMAIRSOLS   18 September 1943.

Subject: COMMENDATION.

To: Strike Command, Aircraft Solomon Islands.

As Commander Aircraft Solomon Islands, I desire to commend you and the Officers and Men of Strike Command, Aircraft Solomon Islands on the results of the special Air Operations carried out so successfully during the period of Sept. 14 to 16 inclusive.

Preceding the special operations, your command had been taxed to the utmost over a long period of strenuous and continuous operations against the enemy in the Solomon Islands.

The further demands on personnel and equipment to perform the additional tasks were accomplished in a superb manner by each and everyone, thereby exhibiting a loyal devotion to duty; all exemplary of the highest traditions of the military service.

To the maintenance personnel for their untiring efforts that made possible the execution of the numerous air attacks, and to the flight crews, who by skillful employment of their weapons dealt telling blows against the enemy air and surface objectives, my praise and congratulations.

It is directed that this commendation be brought to the attention of all members of your command.

/s/ N. F. TWINNING
Major-General, U.S.A.

5. Commendation Message

OPER PRIORITY   SECRET

FROM: AIROPS CHERRY BLOSSOM   020150

TO: AIROPS MUNDA
DOG DAY BOMBING PLANE AND FIGHTING PLANE SUPPORT
SUPERIOR

6. Commendation Message

FROM: CTF 31   020612

TO: COMAIRSOLS

INFO: COMAIRSOPAC, COMSOPAC

YOUR 02343 X GREATEST APPRECIATION AND THANKS FOR
SPLENDID AND EFFECTIVE AIR COVER WHICH I AM
CONVINCED SAVED OUR SHIPS FROM SERIOUS DAMAGE BY
TWO HEAVY STRIKES

NOTE: REF IS PROBABLY CAS'S 012343:

REF: COMAIRSOLS 012343: DETAILS AIR ACTION FIRST ALL
TIMES LOVE X 21 BAKER 24'S HIT KAHILI FROM 19,000 FEET
1000 HOURS CMA DROPPING 125 HALF TONNERS WITH
COVERAGE SOUTHWEST OF RUNWAY AND DISPERSAL AREA X
HEAVY CALIBRE ABLE ABLE:: ETC

7. Commendation Message

PRIORITY   SECRET

FROM: CTF31   020617

TO: COMAIRSOLS

INFO: COMAIRSOPAC COMSOPAC

BOMBING OF BEACH IMMEDIATELY PRIOR LANDING EXCELLENT TIMING AND EXECUTION AND VERY EFFECTIVE IN REDUCING ENEMY RESISTANCE TO ASSAULT WAVES AND THEREBY MINIMIZING OUR ATTACK CASUALTIES

8. **Commendation Message**

PRIORITY   SECRET

FROM: COMSOPAC   020700

INFO: COMGENSOPAC COMGEN FMAC ALL TFC SOPAC CINCPAC COMINCH COMSOWESPAC CINCSOWESPAC

I WISH TO EXTEND MY ADMIRATION AND CONGRATULATIONS TO YOU AND TO ALL HANDS CONCERNED FROM PILOTS TO GROUND CREWS FOR THE SPLENDID AIR EFFORTS PRECEDING AND DURING THE TREASURY AND EMPRESS AUGUSTA BAY INITIAL LANDINGS X YOU SAID YOU COULD NEUTRALIZE THE BOUGAINVILLE AIRFIELDS X YOU DID AND HOW X YOU PROMISED EFFECTIVE AIR COVER AND SUPPORT AND KEPT THAT PROMISE IN OUTSTANDING FASHION X THE JAPS HAVEN'T RECOVERED YET SO KEEP THEM FALLING X HALSEY XX XX

9. **Commendation Message**

ROUTINE   SECRET

FROM: COMAIRSOLS   021111

TO: STRIKE BOMBER

YOUR BLASTING OPERATIONS OF THE AIRFIELDS OF BOUGAINVILLE SET THE STAGE AND MADE POSSIBLE THE

HIGHLY SUCCESSFUL LANDING OPERATIONS COMPLETED TODAY - A THOROUGH JOB WELL DONE XX XX

10. Commendation Message

ROUTINE   SECRET

FROM: COMAIRSOPAC

TO: COMAIRSOLS

INFO: AIRSOL

COMAIRSOPAC PASSES WITH PLEASURE COMSOPAC 020700 FOR ACTION TO WHICH HE SUFFIXES HIS OWN WELL DONE XX XX

11. Commendation Message

TO: AIROPS GUADAL   COMAIRFORCE SOLS

INFO: COMAIRSOPAC   COMGENFMAC

NOVEMBER FIRST OUR SUPPORT BOMBER (S) APPROACHING HIT WITH PRECISION AND GOOD EFFECT X

12. Commendation Message

FROM: COMAIRSOLS

TO: STRIKE COMMAND – FIGHTER COMMAND

TO THE PILOTS AND CREWS WHO PUSHED THROUGH THE BAD WEATHER TODAY COMAIRSOLS SENDS WELL DONE X THE DESTRUCTION AND HAVOC CREATED AT KAHILI AND KARA IS A DEMONSTRATION OF WHAT OUR AIRCRAFT CAN DO WHEN SKILLFULLY OPERATED XX

13. Commendation Message

FROM: HALSEY
TO: COMAIRSOPAC
INFO: COMAIRSOLS COMFAIRWING ONE
           ALL TFC SOPAC

"I AM PROUD OF THE CREWS WHO FLY OUR DAILY SEARCH PLANES X THEIR HABIT OF LOCATING X HARRASSING AND SLAMMING DOWN JAPS OF ALL VARIETIES IS AN INSPIRATION TO ALL HANDS X HALSEY"

14. Commendation Message

From: Air Operations, MUNDA.
To: Air Units, Airops, KOLI, VELLA, RUSSELLS, SEGI.

Subject: Air Op MUNDA, Dispatch 021044 of 2 November 1943.

Subject dispatch is quoted herewith:

THE COMPLETE SUCCESS OF LANDING OUR TROOPS ON BOUGAINVILLE BEACHES WAS DUE TO THE OUTSTANDING PERFORMANCE OF YOUR FIGHTERS WHO KEPT THE RAT FROM THE CHEESE AND TO THE UNTIRING EFFORTS OF BOTH OFFICERS AND MEN RESPONSIBLE FOR OPERATIONS AND MAINTENANCE WHO KEPT THE BALL ROLLING. A TOUGH JOB WELL DONE."

15. Commendation Message

From: Commander-in-Chief, Pacific.
To: Commander South Pacific, for all Task Force
           Commanders, South Pacific.
Subject: CINCPAC Dispatch 020251, dated 2 November 1943.

The subject dispatch is quoted herewith:

"CINCPAC TO COMSOPAC X FOR HALSEY AND HIS TASK
FORCE COMMANDERS X AS USUAL YOUR FORCES HAVE
TURNED IN ANOTHER FINE JOB X WELL DONE X NIMITZ XXX"

16. Commendation Message

From: Commander South Pacific, KOLI.
To: Commander Task Force 38, All Air Units.

Subject: ComSoPacKOLI dispatch No. 020636 dated 2 November
1943.

"WELL DONE ON YOUR FIRST AND SECOND STRIKES X
AS A RESULT OF YOUR BOMBING AND TF 39 BOMBARDMENT
CMA BUKA IS NOT CONTRIBUTING PRESENTLY TO THE JAPS
WAR EFFORT X HALSEY XXX"

17. Commendation Message

<u>SECRET</u>   Day Ending 0800 15 June 1944

<u>THIS DOCUMENT WILL BE DESTROYED BY BURNING AFTER
IT HAS SERVED THE PURPOSE FOR WHICH INTENDED</u>

HAIL AND FAREWELL
ADMIRAL HALSEY'S PARTING MESSAGE
To all SoPac bases and ships Admiral William
Halsey has dispatched the following message:

"Well done" to my victorious all-ser-
vices South Pacific fighting team.
You have met, measured, and mowed
down the best the enemy had on land
and sea and in the air. You have
sent hundreds of Tojo's ships,
thousands of his planes, tens of
thousands of his slippery minions
whence they can never again attack

our flag, nor the flags of our allies.
You beat the Jap in the grim victory
at Guadalcanal, you drove him back
and hunted him out; you broke his
offensive spirit in those smashing
Bougainville-Rabaul blows at his
ships and planes and troops in
November 1943 and you have smeared
him and rolled over him to easily
occupy Emirau. And now, carry on the
smashing South Pacific tradition
under your new commanders, and may
we join up again further along the
road to Tokyo.

HALSEY

18. Commendation Message

ROUTINE   200729/99   PLAIN

IT IS WITH PRIDE AND GRATIFICATION THAT I TRANSMIT
THE FOLLOWING MESSAGE QUOTE FOR ADMIRAL CHESTER W
NIMITZ X AT THE REQUEST OF THE HOUSE OF
REPRESENTATIVES, UNANMOUSLY EXPRESSED TO YOU THE
OFFICERS AND MEN SERVING WITH YOU AND ALSO THE
PARTICIPATING AIR CORPS OUR GRATEFUL THANKS AND
ADMIRATION FOR THE GREAT ACCOMPLISHMENTS DURING
ALL OPERATIONS AND ESPECIALLY THOSE OF THE IMMEDIATE
PAST X WE SALUTE YOU ALL AND SEND WARMEST REGARDS X
SIGNED SAM RAYBURN SPEAKER HOUSE OF
REPRESENTATIVES UNQUOTE.

MY ANSWER IS AS FOLLOWS QUOTE THE HONORABLE SAM
RAYBURN SPEAKER OF THE HOUSE OF REPRESENTATIVES
WASHINGTON D.C. X MY HEART FELT THANKS FOR THE
MESSAGE OF CONGRATULATIONS JUST RECEIVED FROM THE
HOUSE OF REPRESENTATIVES TO THE OFFICERS AND MEN OF

THE UNITED STATES PACIFIC FLEET AND PACIFIC OCEAN
AREAS X WE ARE ENORMOUSLY PROUD TO HAVE THE
CONFIDENCE AND ENCOURAGEMENT OF THE HOUSE X THERE
EXISTS WITH US A KEEN AWARENESS THAT OUR
ACHIEVEMENTS CAN BE TRACED BACK TO THOSE WHO WERE
INSTRUMENTAL IN PROVIDING US WITH THE MEANS TO OUT-
SHOOT "OUTFIGHT" AND OVERPOWER THE ENEMY X THE
VICTORIES WE ARE WINNING IN JAPANESE HOME WATERS
HAD THEIR GENESIS ON MAIN STREET AND IN THE HALLS OF
CONGRESS X WE ARE PROFOUNDLY GRATEFUL FOR THIS
CONTINUED GENEROUS AND ENTHUSIASTIC SUPPORT X YOUR
HEARTFELT MESSAGE WAS TRANSMITTED AT ONCE TO
EVERYONE UNDER MY COMMAND X SIGNED C.W. NIMITZ
"FLEET ADMIRAL" US NAVY" COMMANDER IN CHIEF" PACIFIC
FLEET AND PACIFIC OCEAN AREAS UNQUOTE

ALPOA NO. 99

Appendix E - Ships That Transported CASU-11

The following ships were involved in transporting CASU-11 to the war in the Western Pacific and returning them home. This list only includes the ships that carried the big lifts of several hundred men at one time. As O'Flynn shared, we had "people coming in and people leaving all through the year" on many different ships. In early 1943 the SS President Polk took the men of CASU-11 from San Diego, California to Guadalcanal. In May of 1943, the USS Munargo delivered Lt.j.g. McAteer to Guadalcanal. During the summer of 1944, the USS Rochambeau and the SS Monterey brought CASU-11 home to San Francisco and the USS General Harry Taylor brought ACM Hays home.

On 3 April 1945, the USS Cepheus delivered a small detachment of CASU-11(F) from Hawaii to the beaches of Okinawa. This included seven officers and four enlisted men. During mid-May 1945, the remaining 43 officers and 843 enlisted men arrived on Okinawa. The name of the ship they rode is unknown. On August 1945 the Carrier Aircraft Pool Detachment of CASU(F)-11 riding onboard the USS Burleson arrived at Yonabaru Airfield, Okinawa.

At the end of the war the men of CASU(F)-11 did not leave Yonabaru Airfield en masse. On 1 November 1946, CASU(F)-11 became FASRON-122 and it continued to work at Yonabaru Airfield. Personnel arrived and departed as the needs of the Navy required, and they used the available shipping to and from Okinawa for transportation.

SS President Polk (AP-103)

SS President Polk (AP-103). Navy Photo.

SS President Polk (AP-103) specifications:

"Built in Newport News, VA, by Newport News Shipbuilding & Drydock. Keel laid down 7 October 1941 with launching 28 June 1941.

Deadweight tonnage - 10,500 tons
Length overall - 492 feet
Beam - 69 feet 6 inches
Speed - 18.4 knots
Radius - 14,200 miles

Ship Complement - Officers 30, Enlisted 324
Troop Capacity - Officers 150, Enlisted 2336
Ship's Master - February 1943 – Captain Gregory Cullen

Armament
One 5 inch 38 caliber gun
Four 3 inch 50 caliber dual-purpose* guns
Four Bofors 40 mm gun mounts

*Dual-purpose means can shoot surface targets on sea and land, and can shoot flying aircraft."[1]

On 6 November 1941, the SS President Polk was delivered to American President Lines (APL).[2] On 5 December she began operating as the SS President Polk, a transport under the War Shipping Administration's charter supplying Pacific bases. Sometime during 1942 the troop capacity was expanded from 96 to 2,840 by building bunks within some of the cargo holds. After September 1943 when the ship had been requisitioned and acquired by the Navy as a troop ship, those who rode in SS President Polk across the Pacific had much better accommodations. On 4 October 1943, the ship was commissioned at San Diego as USS President Polk (AP-103).

USS Munargo (AP-20)

USS Munargo (AP-20). Navy Photo.

USS Munargo (AP-20) specifications:

The SS Munargo was "launched on 17 September 1921 and delivered 11 December 1921 by the New York Shipbuilding Corp., Camden, New Jersey to the Munson Steamship Lines.

Tonnage - 7,100 GRT [Gross Registered Tons]
Length - 432 feet
Beam - 57 feet 6 inches
Draft - 23 feet 7 inches
Speed - 16 knots
Radius - 11,000

Ship Complement - 254 crewmen
Troop Capacity - 297 plus

Armament
One 5 inch gun,
Four 3 inch guns,
Eight machine guns"[3]

The USS Munargo had an illustrious past. In 1933 the ship was outbound from the main channel south of Governor's Island in the New York harbor, changed course, and collided with the ship Deutschland.[4]

On 4 June 1941 the ship was commissioned the USS Munargo (AP-20) with Cmdr. Harold F. Ely in command.[5]

USS General Harry Taylor (AP-145)

USS General Harry Taylor (AP-145). Navy Photo.

USS General Harry Taylor (AP-145) specifications:

USS General Harry Taylor was "Launched, 10 October 1943 and on 29 March 1944 was acquired by the Navy.

Displacement - 9,950 tons
Length - 522 feet 10 inches
Beam - 71 feet 6 inches
Draft - 26 feet 6 inches
Speed - 16.5 knots (trial)

Crew - Officers 32, Enlisted 324
Troop Capacity - Officers 228, Enlisted 3,595

Armament
Four single 5 inch 38 caliber dual purpose gun mounts
Four twin 1.1 inch gun mounts (replaced by four twin 40mm AA (anti-aircraft) gun mounts)
Fifteen twin 20mm AA gun mounts"[6]

"On 8 May 1944 the ship was commissioned by the Navy as the USS General Harry Taylor (AP-145) with Capt. James L. Wyatt in command. Following a shakedown cruise off San Diego, the General Harry Taylor sailed on 23 Jun 1944 from San Francisco for Milne Bay, New Guinea with troop reinforcements. After returning to San Francisco on 3 August 1944 with veterans of the Guadalcanal campaign embarked, she continued transport voyages between San Francisco and the island bases in the Western Pacific. During the next 10 months she steamed to New Guinea, the Solomons, New Caledonia, the Marianas, the New Hebrides, the Palaus, and the Philippines, carrying troops and supplies for America's vast amphibious sweep across the ocean to Japan. On 29 Jun 1945 she departed San Francisco for duty in the Atlantic."[7]

USS Rochambeau (AP-63)

USS Rochambeau (AP-63). Navy Photo.

USS Rochambeau (AP-63) specifications:

"Built in 1933 by Societe Provencals de Construction Navales at La Ciotat, France under name: MARECHAL JOFFRE.

Deadweight tonnage - 14,242 tons
Length overall - 470 feet 10 inches
Beam - 63 feet 11 inches
Speed - 15 knots"[8]
"Radius - 9,404 miles

Troop Capacity - 3,015 passengers"[9]

Armament
One single 5 inch 38 dual purpose gun mount
Four single 3 inch (76 mm) dual purpose gun mounts
Eight single 20 mm gun mounts

The USS Rochambeau "was at Manila 7 December 1941 and was requisitioned there by the Allies and taken under protective custody by the U.S. Navy. The Navy put a crew on her and sent her under escort on 18

December 1941 to Balikpapan and then to Darwin, Australia, where she was attacked by torpedoes on 23 January 1942 just before her departure to Sydney. By special arrangement the ship then hoisted the U.S. flag and departed Sydney for the United States on 24 March 1942 under the command of an American naval officer.

The vessel was commissioned as USS ROCHAMBEAU on 27 April 1942 and was converted by Moore Dry Dock Co., Oakland, California, between 27 April and 28 September 1942."[10] "During the period October 1942 and January 1945 the Rochambeau made ten round trips from California ports to the Southwest Pacific."[11] The last day of June 1944 the Rochambeau picked up an unknown number of CASU-11 personnel for transport back to the USA. This began the eventual, total removal, of all CASU-11 personnel from Guadalcanal.

SS Monterey

SS Monterey. Navy Photo.

SS Monterey specifications:

On 10 October 1931, "SS Monterey, a luxury ocean liner, was launched. The ship was completed during April 1932. Monterey was the third of the four ships of the Matson Lines 'White Fleet', which were designed by William Francis Gibbs. The others included the SS Malolo, SS Mariposa and SS Lurline. Monterey was identical to Mariposa and very similar to Lurline. During World War II Monterey was used as a troopship operated by Matson as agents of the War Shipping Administration (WSA). Monterey was a large, fast transport capable of sailing independently and was allocated to serving Army troop transport requirements."[12]

"Built in 1932 by Fore River Shipyard Bethlehem Shipbuilding Corporation.

Deadweight tonnage - 18,017 tons
Length overall - 632 feet
Beam - 79 feet
Draft - 28 feet 2 inches
Speed - 20.5 knots
Radius - 18,000 miles

Ship Complement - 360

Troop Capacity - 3,500
Armament
None

On 3 July 1944, the ship departed San Francisco and steamed to Honolulu, Milne Bay, Oro Bay and Guadalcanal."[13] During the 23 July trip from Milne Bay to Oro Bay, along the northeast coast of Papua, New Guinea, the ship ran aground off Cape Ward Hunt (08-56S,149-16E), troops were offloaded and the ship was refloated with the 25 July tide. Damage was slight. Early August the Monterey's last South Pacific stop was Guadalcanal where the last members of CASU-11 boarded and departed Guadalcanal for San Francisco.

USS Cepheus (AKA-18)

USS Cepheus (AKA-18). Navy Photo.

USS Cepheus (AKA-18) specifications:

"USS Cepheus (AKA-18) was an Andromeda-class attack cargo ship named after the northern constellation Cepheus. She was one of a handful of United States Navy AKAs that were manned by Coast Guard officers and crew during World War II. She served as a commissioned ship for 2 years and 5 months. She was launched 23 October 1943 by the Federal Shipbuilding and Drydock Company, Kearny, New Jersey."[14]

Deadweight tonnage - 6,556 tons
Length overall - 459 feet
Beam - 63 feet
Speed - 16.5 knots
Radius - Unknown

Troop Capacity - Officers 44, Enlisted 360

Boats
Eight LCMs [Landing Craft Mechanized]
One LCP(L) [Landing Craft Personnel, Large]
Fifteen LCVPs [Landing Craft Vehicle, Personnel]

Armament
One 5 inch 38 caliber dual purpose gun.

Four 3 inch guns.
Eighteen 20mm AA gun mounts."[15]

From March through April 1945 the USS Cepheus transported the 11 man advanced echelon of CASU(F)-11 from Honolulu to Okinawa, Japan.

USS Burleson (APA-67)

USS Burleson (APA – 67). Navy Photo.

USS Burleson (APA-67) specifications:

"USS Burleson (APA-67) a Gilliam-class attack transport, was the only ship of the United States Navy to be named for Burleson County, Texas. Burleson County is located in East Central Texas about 60 miles east of Austin. Her keel was laid down on 22 April 1944 at Wilmington, California, by the Consolidated Steel Corporation under a Maritime Commission contract (MC Hull 1860). She was launched on 11 July 1944 sponsored by Mrs. Darryl F. Zanuck, delivered to the Navy on 7 November 1944, and commissioned on 8 November 1944 with Lieutenant Commander B. Hartley, USNR, in command."[16]

"Displacement - 4,247 tons
Length - 426 feet
Beam - 58 feet
Draft - 16.9 feet
Speed - 6.9 knots.

Ship Complement - Officers 27, Enlisted 293
Troop Capacity - Officers 47, Enlisted 802

Armament

One single 5 inch 38 caliber dual purpose gun mount
Four twin 40mm AA gun mounts
Ten single 20mm AA gun mounts"[17]

"Burleson entered the transport area off Okinawa on the morning of 1 April 1945, the day of the initial assault, but she did not begin unloading until the following day. Those operations continued until 7 April when she put to sea bound for Guam. The attack transport spent the night of 11 April and 12 April at Apra Harbor, Guam, and then resumed her voyage. She arrived in Pearl Harbor on 23 April. Burleson remained in the Hawaiian Islands for two months. She conducted several amphibious training exercises at Maui and underwent repairs at the Pearl Harbor Navy Yard.

On 25 June 1945, the ship got underway from Pearl Harbor on her way back to the Western Pacific."[18] On board were 181 men bound for Okinawa and the Carrier Aircraft Pool Detachment of CASU (F)-11 at Yonabaru Airfield. "After stops at Eniwetok and Ulithi, the attack transport arrived back at Okinawa on 5 August. She unloaded her cargo and disembarked her passengers at Buckner Bay. Burleson stayed at Okinawa beyond the end of hostilities on 15 August and through the end of August."[19]

Appendix F - Airplanes Serviced by CASU-11

CASU-11 received, maintained and repaired numerous types of aircraft. They are listed in chronological order. The sources for the following were the "Locations of U. S. Naval Aircraft, World War II"[1] reports held at the Navy History and Heritage Command and the service records of the CASU-11 aviation rated enlisted members. The notes for this appendix list several publications used to gather specific details about these airplanes.

F4F-3 Wildcat

Grumman F4F-3 Wildcat Fighter. Navy Photo.

The Grumman F4F-3 Wildcat was the precursor to the F4F-4 with less armament, less protection, and non-folding wings. It was built in 1940 as a single-seat aircraft and was soon overtaken by other, more capable, aircraft. It was 28 feet 9 inches long with a wingspan of 38 feet and a maximum takeoff weight of 5,876 pounds. The Wildcat was capable of 325 miles per hour, a ceiling of 37,500 feet, and a range of 845 miles. This aircraft mounted six guns and a small bomb load of 200 pounds.

F4F-4 Wildcat

Grumman F4F-4 Wildcat Fighter. Navy Photo.

The Wildcat was a mid-wing monoplane with only one seat. It was mostly a carrier-based fighter that brought great maneuverability to the U.S. aviation fleet. Aircraft length was 28 feet 9 inches with a wingspan of 38 feet and a maximum takeoff weight of 7,952 pounds. This version of the Wildcat had a compound angle folding wing that folded flat alongside the fuselage. This permitted an aircraft carrier to carry more of these airplanes. The Wildcat was capable of 318 miles per hour, a ceiling of 34,900 feet, and a range of 770 miles. Carrying six machine guns, this was almost totally a fighter as opposed to being a bomber. The Wildcat only carried 200 pounds of bombs.

F6F-3 Hellcat

Grumman F6F-3 Hellcats. Navy Photo.

The F6F-3 Hellcat was a U.S. Navy fighter aircraft designed to replace the F4F Wildcat. The Hellcat, a carrier-based plane, could attain a top speed of 380 miles an hour, could reach as high as 37,398 feet, and could go 945 miles without landing. Aircraft length was 33 feet 7 inches with a wingspan of 42 feet 10 inches and a maximum takeoff weight of 15,413 pounds. The crew only consisted of a pilot. This aircraft was capable of taking considerable fight to the enemy by employing six forward shooting machine guns. This aircraft represented the high point of American carrier-based aviation during the last two years of the war.

SBD-5 Dauntless

Douglas SBD-5 Dauntless. Navy Photo.

The Douglas SBD-5 Dauntless was a scout plane and dive bomber. The Dauntless carried two men sitting under bulletproof glass. When fully loaded with bombs, the Dauntless could fly at 255 miles per hour, had a range of 1,115 miles and had a maximum altitude of 26,850 feet. Aircraft length was 33 feet 1 inch with a wingspan of 41 feet 6 inches and a maximum takeoff weight of 10,700 pounds. Four machine guns and a bomb load capacity of 1,200 pounds made this aircraft a formidable weapon.

During the critical period of April through November 1943, this aircraft was crucial in defending the islands the Allies had taken and needed to take to reach Rabaul. Pilots loved this aircraft. The Dauntless was tough, ready to take a beating, and still get its crew home.

TBF-1 Avenger

Grumman TBF-1 Avenger dropping torpedo. Navy Photo.

The TBF-1 Avenger was the U.S. Navy's main torpedo bomber. It was a mid-wing monoplane with seats for three men. There was a gun turret on top behind the cockpit. Under the fuselage there was another gun to protect the rear. The Avenger could carry a torpedo that weighed almost a ton or it could carry a ton of bombs within the body of the aircraft. Aircraft length was 40 feet 11 inches, wingspan was 54 feet 2 inches and maximum takeoff weight was 17,895 pounds. Maximum speed was 276 miles per hour, maximum ceiling was 22,000 feet, and maximum range was 1,010 miles.

F4U-1 Corsair

Chance Vought F4U-1 Corsair. Navy Photo.

The Corsair F4U-1 was a carrier-based fighter aircraft that had one of the most powerful engines ever carried by a single seat plane. It carried a Pratt and Whitney air-cooled engine with 2,250 horsepower to reach a top speed of 446 miles per hour. Aircraft length was 33 feet 8 inches with a wingspan of 40 feet 11 inches and a maximum takeoff weight of 14,670 pounds. Equipped with six machine guns and capable of carrying 2,000 pounds of bombs, the Corsair F4U destroyed 2,140 enemy aircraft in combat, compared to only 189 losses, for a win ratio of 11:1.

TBM-1C Avenger

Grumman TBM-1C Avenger. Navy Photo.

The TBM-1C Avenger was a torpedo bomber with a crew of three aviators that operated from both aircraft carriers and remote island-based airfields. The Avengers maximum speed was 276 miles per hour, with a maximum ceiling of 23,400 feet, and a maximum range of 1,010 miles. Aircraft length was 40 feet 11 inches with a wingspan of 54 feet 2 inches and a maximum takeoff weight of 17,895 pounds. This aircraft was very capable with a capacity for carrying a 2,000 pound torpedo, or four 500 pound bombs, or a 250 gallon auxiliary gas tank. It also carried three machine guns.

SBD-4 Dauntless

Douglas SBD-4 Dauntless. Navy Photo.

The SBD-4 Dauntless was a two seat dive bomber and the precursor to SBD-5 Dauntless. Seven hundred and eighty SBD-4s were delivered to the American fighting forces. This aircraft mounted four guns, two forward and two rear-facing and had a bomb capacity of 2,250 pounds. Aircraft length was 33 feet 1 inch with a wingspan of 41 feet 6 inches and a maximum takeoff weight of 10,700 pounds. Maximum speed was 255 miles per hour with a maximum range of 1,115 miles and maximum ceiling of 25,530 feet.

SBC2C-1 Helldiver

Curtiss SB2C-1 Helldiver. Navy Photo.

The Curtiss SB2C-1 was a carrier-based dive bomber. The Helldiver was used as a scout and bomber aircraft. It was two-seated, and it could fly at 295 miles an hour, climb to 25,400 feet and travel 1,235 miles. Aircraft length was 36 feet 8 inches with a wingspan of 49 feet 9 inches and a maximum takeoff weight of 16,616 pounds. This aircraft was loaded with two cannons, two machine guns and could carry 2000 pounds of bombs.

SOC-3 Seagull

Curtiss SOC-3 Seagull. Navy Photo.

The SOC-3 Seagull was a two seat scout observation biplane which performed aerial reconnaissance. After locating enemy forces this aircraft worked as a gunfire spotter for Allied battleships. The pilot passed the target location, advised whether the big guns were hitting the target, and provided corrections to bring gunfire onto the target. It had a maximum speed of 165 miles per hour, maximum range of 675 miles and maximum ceiling 14,900 feet. Aircraft length was 26 feet 6 inches with wingspan of 36 feet and a maximum takeoff weight of 5,437 pounds. Guns were mounted one forward and one pointed rearward. Bomb load was 650 pounds.

FG-1 Corsair

Goodyear FG-1 Corsair. Navy Photo.

To increase the speed of production, Goodyear was added as a contractor to build Corsairs. Their single seat aircraft carried the FG designation. The engine for this fighter-bomber was a 2,100-horsepower R-2800-18W which permitted a maximum speed of 446 miles per hour. Maximum range was 1,005 miles with a maximum ceiling of 41,500 feet. Starting in World War II and continuing through 1952, 1,074 Corsair FG-1's were delivered to the Navy and Marine Corps. The airplane was equipped with six machine guns and capable of carrying two 1,000-pound bombs.

FM-2 Wildcat

Eastern Aircraft FM-2 Wildcat. Navy Photo.

Eastern and Grumman aircraft companies collaborated on the new FM-2 Wildcat design. This was required by the Navy for its "Jeep" or Escort Carriers. The single seat FM-2 was the most numerous and best performing Wildcat with 1,350 horsepower, bigger vertical tail surfaces and more fuel capacity, earning the name the "Wilder Wildcat." This enabled the FM-2 to fly 320 miles per hour to a maximum range of 830 miles with a maximum ceiling of 34,000 feet. The FM-2 had four wing guns versus six in the original Wildcat. Removing those two guns allowed the plane to carry more ammunition to the fight.

TBM-3 Avenger

TBM-3 Avenger. Navy Photo.

Starting in mid-1944, the TBM-3 Avenger began production with a more powerful powerplant and wing hard points for drop tanks and rockets. The three seat dash-3 was the most numerous of the Avengers with about 4,600 produced. Besides the traditional surface role (torpedoing surface ships), Avengers claimed about 30 submarine kills, including the cargo submarine I-52. They were one of the most effective sub-killers in the Pacific theatre, as well as in the Atlantic, when escort carriers were finally available to escort Allied convoys. Maximum speed was 276 miles per hour with a range of 1,010 miles and a maximum ceiling of 30,100 feet. Aircraft length was 40 feet 11 inches, a wingspan of 54 feet 2 inches and a maximum takeoff weight of 17,895 pounds. This aircraft carried three machine guns, up to eight 3.5 inch forward-firing rockets and up to 2,000 pounds of bombs or one 2,000-pound torpedo.

F6F-5 Hellcat

Grumman F6F-5 Hellcat. Navy Photo.

The Grumman F6F Hellcat was a single seat carrier-based fighter aircraft designed to replace the earlier F4F Wildcat and to counter the Japanese Mitsubishi A6M Zero. It was the United States Navy's dominant fighter in the second half of the Pacific War.

Powered by a 2,000 horsepower Pratt & Whitney R-2800 Double Wasp engine, the F6F was an entirely new design, but it still resembled the Wildcat in many ways. Aircraft length was 33 feet 7 inches with a wingspan of 42 feet 10 inches and a maximum takeoff weight of 14,250.

The F6F was best known for its role as a rugged, well-designed carrier fighter which was able, after its combat debut in September 1943, to outperform the A6M Zero and help secure air superiority over the Pacific theater. A total of 12,275 were built in just over two years. Maximum speed was 376 miles per hour with a range of 1090 miles and a maximum ceiling of 37,300 feet. The aircraft was equipped with six guns or two cannon, and could carry up to 4,000 pounds of bombs and a single torpedo, mounted on centerline and wing racks.

311

PB4Y-2 Privateer

PB4Y – 2 Privateer. Navy Photo.

The Consolidated PB4Y-2 Privateer was a Navy World War II patrol bomber derived from the Consolidated B-24 Liberator. An aircraft meeting specific Navy needs was desired, and Consolidated developed a dedicated long-range patrol bomber in 1943, designated PB4Y-2 Privateer. Aircraft length was 74 feet 7 inches with a wingspan of 110 feet and a maximum takeoff weight of 65,000 pounds. The crew had 11 members: two pilots, a navigator, a bombardier, five gunners and two radio operators.

Defensive armament on the PB4Y-2 was increased to twelve .50-in (12.7 mm) M2 Browning machine guns in six power operated turrets (two dorsal, two waist, nose and tail). Offensively the Privateer could carried up to 12,800 pounds of bombs, mines, or torpedoes. The Navy eventually took delivery of 739 Privateers, the majority after the end of the war.

Appendix G - SBD Mission Planning and Flying

CASU-11 was an airplane maintenance and repair unit, and its work was intimately related to the mission planners and pilots. Once the mission was planned, the number and types of aircraft were selected and pilots were assigned to them. This air plan was published, and the CASU took action to ready the selected airplanes for takeoff. These actions could include mounting and fueling extra fuel tanks, loading the proper ordnance, and arming the gun magazines with the correct size bullets. After pilots had properly prepared aircraft, then it was time to execute the mission.

As a general rule Strike Command aircraft, SBD's and TBF's almost always had Fighter Command aircraft as escort and protection from enemy fighters. If weather prevented the Strike aircraft from finding their Fighter Command protection then the mission was almost always cancelled.

The following is an excerpt from Strike Command, Aircraft Solomons *War Diary for the period 15 March 1944 to 1 June 1944* dated 15 July 1944.[1]

"INTELLIGENCE
STRIKE COMMAND
COMMAND AIRCRAFT SOLOMONS

CONFIDENTIAL   1 May 1944

FACTS OF A TYPICAL STRIKE

HOW IT ALL STARTS:

ComAirSols (Commander of Aircraft in the Solomon Islands) decides, on the basis of reports from various intelligence sources, and from directives received from higher authority, that it is necessary to bomb a target. ComAirSols so informs Strike Command at the morning conference and outlines the purpose of the attack. Strike Command then decides how many planes are needed to accomplish that purpose, what type of bombs they should carry, and the time of the attack. The decisions reached are based on weather forecasts, availability of aircraft, effectiveness of different types of bombs and fusings, number and location of anti-aircraft, size and type of targets, coordination with other missions of the day, the desirability of varying the attack and, as far as possible, considerations from the pilots' point of view, such as early return to the field, not too many missions on consecutive days, etc. Strike Command has SBD's, TBF's, and PV's under its control; it also works in close cooperation with Bomber Command, which has B-25's and B-24's, and with Fighter Command which has P-38's, P-39's, P-40's, F6F's and F4U's.

HOW THE PILOTS "GET THE WORD":

After the respective squadrons have been informed by Strike Command Operations as to how many planes they will be expected to provide for the strike, schedules are drawn up and the pilots who are to go on the strike report to Strike Command for a briefing (usually the afternoon before the strike, if takeoff is to be in the morning). An hour before the briefing, pilots drop around to the Briefing Hut and carefully study photos of the target assigned to each pilot and the surrounding area.

At the briefing, Strike Intelligence officers, who have made a special detailed study of the target, lecture on:

The approaches to the target,
The appearance of the target,
The location of the target,
The defenses around the target,
The interception to be expected, and

The necessary survival information (location of coastwatcher, friendly natives, winds and currents, crash boats, "Dumbos," etc.)

Recent aerial photographs, accurate maps, reliefs and large sketches of the area are used for purposes of illustration.

The Operations Officer then details the tactics to be used. He discusses the takeoff, rendezvous, bomb loadings and fusings, the manner of approach to the target, the specific target for each division or section, the route of retirement and place of rally, the route home, and fighter escort.

HOW THE PILOT GETS READY TO TAKEOFF:

Most pilots carry the following equipment on all strikes:

> An emergency medical kit,
> A pistol, loaded, with extra ammunition (PARTICULARLY TRACER AMMUNITION),
> A sharp knife, in a position convenient to the pilot's hands, should he have to cut shroud lines under water.
> A small mirror for signaling,
> A pocket compass, in a waterproof container.

All of the above items should be so secured to the pilot's person that there is no danger of them breaking loose if he has to use his parachute.

With his "survival equipment," parachute, flight gear, and plotting board loading him down, the pilot "staggers" into the SBD or TDF Operations Hut about an hour before the scheduled time of takeoff to look at the schedule board. There he finds posted his plane number, his radio call, and the identification signals and code of the day. He then proceeds to his plane.

At the plane, the pilot, plane captain or ordnance man carefully inspects the bombs (and bomb bays on TBF's) to see that all bombs are loaded and fused properly, and MOST IMPORTANT that the arming wires are properly installed. Dropping a dud, and risking your life to do it, is very unsatisfactory; often it can be prevented by proper inspection of the bombs and bay.

The pilot should make the usual routine inspection of his plane, and the gunners should check their guns.

The radios have been tested by the radio mechanic and, for purposes of radio silence, no transmission checks are made.

Prior to the takeoff, the pilot passes on to his crew any of the information gained at the briefing which the latter needs to know to carry their portions of the day's work. The question of strafing (whether and where to strafe) usually is discussed.

TAKEOFF:

Usually the SBD's takeoff first, the TBF's follow. On most of the strips in use in this area, it is the general practice to takeoff in two-plane sections. Each section starts rolling as soon as the preceding section is airborne. Their takeoff is regulated by a signal officer on the edge of the strip.

A "carrier join-up" usually is effected. The leading planes climb straight ahead after takeoff with a moderate power setting, for a length of time which is dependent upon the number of planes in the formation (about 4 minutes for 18 planes, 6 minutes for 24). The leading plane then makes a gentle 180 degree turn, and all the planes should be joined up by the time the leading plane is back even with the runway.

FORMATION:

Flying six-plane divisions in three-plane sections, the SBD flight has its sections either stepped down or up, depending on squadron doctrine. The flight is composed of diamonds, of two divisions each. If the approach to the target is to involve a turn to starboard, the division leading and making up half the diamond (see sketch) flies its sections in positions A & C.

```
            X
           (A)
         X     X


       X             X
      (C)           (B)
       X    X     X    X


            X


       X  (D)  X
```

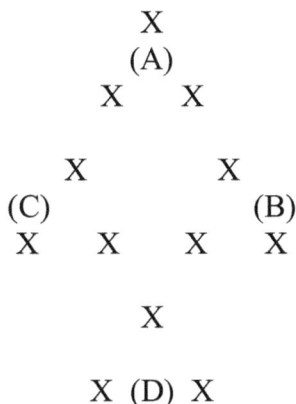

The second division then takes positions B & D, completing the diamond. Thus, the flight leader and the leader of the other division can be followed easily, while at the same time the following section leaders of each division obtain a clear view of the target as the swing is made.

When the turn is to be to port, the leading SBD division takes positions A & B and the second division C & D so that the turn is accomplished efficiently.

The third and fourth divisions form a second diamond arranged in exactly the same manner, and if there is a fifth or sixth division, a third diamond is flown, and so on.

Combined with the above advantages, this closely knit pattern makes attack by enemy planes less attractive and cover by friendly fighters easily maintained.

Cruising formation (intervals as great as 100 feet between planes, if necessary for comfort) generally is flown in safe areas. Close formation is flown until the rendezvous is completed. The signal to open out (swishing the tail) is then given and loose formation is flown until within 40 or 50 miles of the target. The signal for closing up is wiggling the flippers.

The spares (or reserves) fly a little off to one side and a little above the rest of the formation. Whenever a plane turns back, his absence will be obvious by the irregularity of the formation. The next spare in the line then uses his altitude advantage to catch up to complete the division, pulls up beside the vacancy and slides in.

On strikes, SBD's, flying in a column of diamonds, are ahead of the column of TBF 12-plane formations.

TBF's are flown, according to squadron doctrine, in two types of six-plane divisions. One comprises three plane sections, stepped down:

<pre>
            X

        X   X

            X

        X   X
</pre>

The sections follow so closely, one on the other, that the resulting spearhead makes a formation that can be protected easily. The sections maintain the same positions whether the turns are to port or starboard.

The other type of division is composed of three two-plane sections, stepped either down or up:

<pre>
            X

            X

        X   X

      X   X
</pre>

While not as closely knit as the first type, this formation is very maneuverable, affording each pilot an excellent view of the target and other planes while making turns and peel-offs. Turns to port or starboard do not affect the positions of the sections.

Because Strike Command often has squadrons assigned for tours of duty at the same time that fly different formations, a combination formation that would not entail the adoption of new flight practices for some squadrons is employed.

This is done by placing a division of three-plane sections inside the wedge formed by a division of two-plane sections. The formation is close and very maneuverable:

```
          X

          X

       X  X  X

     X  X  X  X

          X

       X     X
```

On strikes, TBF's, flying in a column of either of these formations, are behind the column of SBD diamonds.

RADIO DISCIPLINE:

Strike Command planes do not maintain silence, but they DO observe RADIO DISCIPLINE. That is, the radio is used only in an emergency, (when a plane is in trouble), to give a message that is absolutely essential to the proper execution of the attack, or to report the sighting of an enemy ship.

Strike Command planes are on the same frequency as the escort fighter planes; and unless the radio discipline is maintained strictly, the air will be filled with a lot of useless chatter, which will prevent the fighter escort planes from getting the call for help when a Zeke or Oscar gets on your tail. Also, the Japs are often on the same frequency as that of the Strike Command planes, and the little Nips can get plenty of valuable information from your idle talk. In this connection, be sure that the plane's radio system is on the intercommunications hookup before you start talking to your crew. In short, don't worry about Bill's running lights being on or how much water the gunner has. KEEP THE LIPS ZIPPERED.

ATTACK:

Enroute to the target, the formation will climb to between 13,000 and 15,000 feet. The TBF's usually remain about a half-mile behind, and slightly below the SBD's. Just prior to entering the target area (that is, the area in which AAA fire will be encountered), the high-speed approach is started. At this point pilots-especially SBD pilots-should try to determine the location of their specific targets by actual visual sighting.

The attack break-up is affected by a rapid series of peel-offs toward the target. This may involve simultaneous peel-offs by individual planes, sections or even divisions - with SBD's leading the attack, except in a few instances where a division of SBD's followed the TBF's to take care of any Jap AAA gunners that might have come out of their holes to take a few shots at the TBF's. The approach, evasive tactics, and dives of individual planes depend almost entirely upon the circumstances, the training of the squadrons and preferences of flight leaders. Final pushover is usually at around 8,000 feet. Pilots MUST pick up their target and keep their eyes on it from this point on. Speed varies during the dive from 200-275 knots up (TBF's are red-lined at 316 knots). Pilots have been killed as a result of wing failures, probably due to excessive speeds or jerky use of the controls at high speeds.

Pilots should know thoroughly the diving check-off list, and should check all items three times before diving.

Items to be checked by SBD pilots:

> Trim,
> Mixture - low blower, mixture rich, air intake cold,
> Pitch - 2,000 revs.
> Fuel - either starboard or port main tank.
> Flaps,
> Instruments - cage, if necessary
> Engine - cowl gills closed,

Master bomb switch on,
Gun sight switch on,
Front gun switches on,
Bomb release lever in salvo position,
Cockpit hatch opened,
Camera switch on during dive,
Press electric bomb release button (if used),
PULL MANUAL EMERGENCY BOMB RELEASE.

Items to be checked by TBF pilots:

Radio man should arm the bombs as soon as the high speed approach is started and notify the pilot that "the bombs are armed,"
Pilot should turn on the master arming switch,
Pilot should check the bomb-torpedo selector switch,
Pilot should turn on the wing gun switch (when these three switches are in the proper position, all three will be pointing towards the pilot),
Cockpit hatches open (they will burst if they are left closed during a dive),
Supercharger on low blower, mixture full rich,
Bomb bays open (do not open bomb bays until you are well clear of the formation; and as an added precaution against AA fire do not open bomb bay doors until ready to release bombs).

MOST IMPORTANT OF ALL: Pull your emergency bomb releases immediately after pushing your electric release.

The TBF check off is worth repeating:

Bombs armed,
Master arming switch on,
Bomb-torpedo switch on bombs,
Gun switch on,
Hatches open,
Low blower,
Mixture rich,
Bomb bays open,
Press electric bomb release,

PULL EMERGENCY RELEASE!

The importance of pulling the emergency release, and of yanking it as hard as possible, cannot be over emphasized. A former Operations Officer of Strike Command maintained that a TBF pilot whose bombs failed to release had no justifiable excuse unless he brought the emergency release back in his hand.

The altitude of release depends upon the type of bomb being dropped. If a 2000-pound bomb is being dropped (or if any other plane in the formation is dropping a bomb that large), planes should drop between 2500 and 2000 feet (in no case below 2000). Planes should be fully recovered by 1500 feet.

Recovery from a dive will usually consist of shallow pullout to clear the bomb blast, followed immediately by a push-over to gain speed and get to the rally point.

ALTITUDES OF ATTACK:

The following table illustrated the altitude which Strike Command has found most effective against land and shipping targets in the South Pacific Theater.

SBD's:

Start of High Speed Approach

(200 Knots or better)	PUSHOVER	RELEASE	PULLOUT
15,000 FEET	8,000 ft.	2,000 ft.	1,500 ft. to 2,500 ft.

Angle of Glide: 70 degrees.

TBF's:

Start of High Speed Approach

(250-300 Knots)	PUSHOVER	RELEASE	PULLOUT
15,000 FEET	8,000 ft.	2,000 ft.	1,500 ft. to 2,500 ft.

Angle of Glide: 30 degrees.

THE RALLY:

The primary purpose of the rally is to get joined up for mutual protection as soon as possible and, also as soon as possible TO GET THE HELL OUT OF THERE. The importance of joining up cannot be overemphasized. Stragglers are mincemeat for enemy fighters. Don't look for your section leader, don't look for your division leader, and don't look for your friend Joe. Just look for the nearest SBD or TBF and join up! Join on Harry, join on Henry, or join up on Ezekiel - BUT JOIN UP, and stay joined up. But, do not push the throttle to the fire wall and try to be first at the rally point, regardless of your position in the attack; it will foul up the rally. If everybody is on the ball, a rally turn will be unnecessary and the formation will be able to proceed directly home. That will reduce by three minutes the length of time spent near the enemy fighters and may well save some lives. On the other hand, if the leader - or pilots reaching rendezvous first - applies full throttle, the formation will never be able to rally. Frequently, to facilitate the rally, the SBD flight leader drops his wheels or switches on his lights to identify himself.

GETTING HOME:

On the way home, pilots should assure themselves that their bombs have been released and that the TBF bomb bays are empty. Bombs often hang up on the shackles or drop from the shackles and remain in the TBF bomb bays. The bays should be checked by the radioman looking into the bay with the Aldis lamp. If there still is any doubt the pilot should slide out of formation, and make sure that he is clear of all planes, open his bay doors, and give the emergency bomb release a vigorous working over. Hung bombs on both SBD's and TBF's should be jettisoned unarmed over the water while out of formation and away from other planes, so as not to endanger your mates. Pilots also should keep a sharp lookout for any of their companions who may be in trouble as the result of AAA damage.

After the formation is out of the danger area, it is usually permissible to go into cruising formation, but DON'T straggle. The formation should get home as a smart, well-dressed flight, and execute a snappy break-up and landing. Nothing makes pilots madder than long delays in landing after sitting on a poorly padded parachute for several hours.

Signed
E.D. MOORE
Lieut. USNR
Intelligence Officer."

Hopefully, you now have a better understanding of what a pilot does in preparing to climb into an airplane that CASU-11 repaired after damage from a previous combat mission and made ready for another go. After doing all of the detailed pre-flight preparations, let's explore how to make a successful bomb run in a SBD Dauntless:

A dive bomber is an aircraft that dives directly at its targets in order to provide greater accuracy for the bomb it drops. Diving towards the target simplifies the bomb's trajectory and allows the pilot to keep visual contact throughout the bomb run. This results in increased accuracy of the bomb drop.

A dive bomber rapidly descends at a steep angle, about 70 degrees in the SBD and requires an abrupt pull-up after releasing its bombs to safely exit the target area. This puts great strains on both pilot and aircraft. It demands an aircraft of strong construction or with some means to slow its dive. The Douglas SBD Dauntless was designed with these attributes.

An aircraft diving vertically minimizes its horizontal motion. When the bomb is dropped, the force of gravity increases the bombs speed along this nearly vertical path. The bomb travels a virtually straight line between release and impact, eliminating the need for the pilot to lead the target. Only wind will affect the bomb after release. However because of the steep delivery and the bomb's heavy weight wind will have only slight affect.

After many trials it was learned dive bombing was the most accurate method for attacking targets, like runways and ships. However, the forces generated when the aircraft pulls out of a dive are huge. One way to manage the impact of this dive maneuver was to restrict dive speed by special flaps.

Dive flaps, also called dive brakes, were employed on the Dauntless SBD dive bomber to create drag, which slowed the aircraft in its dive and provided increased accuracy. These flaps are perforated, split dive flaps, one went up, the other down, positioned on each wing. When in a dive at almost 300 miles per hour with the dive flaps opened it felt as if you were hanging in the air. The flaps held you back from further acceleration staying at a safe speed for the air frame. It was still a very bumpy and noisy ride as 300 miles per hour air passed through the holes in the dive flaps.

During the dive you must keep your eyes on your height indicator, the altimeter, and the bomb sight, mounted in the windshield. You point the plane to aim the bomb at the target. The bombs are released between 2,500 and 1,500 feet above the target and then the pilot must pull back hard on the control stick to pull out of the dive and get away from enemy gunfire as quickly as possible. You must remember to close the dive flaps as you pull out otherwise you would lose a lot of speed when you went from near vertical dive to flying horizontal. You must do three things almost simultaneously, release bomb, pull out of dive, and close the dive flaps. If done correctly you can save some of the dive speed and use it to leave the target behind faster than just engine propulsion. If you forget to close the flaps it is like leaving the brakes on and it seems to take forever to get away from the enemy.[2]

Appendix H - After Action Report on CASUs

The CASU Model was conceived and hastily implemented to meet the needs of a wartime Navy. On 17 June 1944, shortly before CASU-11 arrived home from Guadalcanal the Chief of Naval Operations (CNO) issued a letter noting, "In the past, aircraft maintenance and repair personnel, particularly those attached to CASU's, have been sent overseas inadequately trained."[1] CASU-11 was one of these inadequately trained units when it arrived on Guadalcanal. However, with superb leadership, technical ingenuity, and an abundance of repair parts CASU-11 overcame its training deficiencies. Taking CASU-11's experience into account, the CNO's directive included the following action items:

Ten percent of the most experienced personnel from a returning CASU would transfer to a newly formed CASU as it was getting ready to go overseas. The bulk of the remaining men for this new CASU would come from continental CASUs where they had gained months of experience since they departed basic aircraft maintenance school. All would go through a new school at ACORNTraDet, Port Hueneme, California using the tools and equipment required to maintain the current fleet of aircraft under simulated field conditions of the Western Pacific.

CASU-11 went through this new training process at Port Hueneme, NAF Thermal, and Point Magu prior to going overseas to Okinawa. Some of their experienced hands were retained for this newly reformed CASU(F)-11 and retrained at Port Hueneme. New hands were brought in from the continental CASUs and other returning CASU personnel were gathered and distributed to these reforming units. As part of this process CASU(F)-11 shifted from working on fighter aircraft to long-range, multi-engine aircraft. As well, CASU(F)-11 grew to 44 officers and 844 enlisted men. This was much larger than the Guadalcanal unit.

It is not easy to write a report card on the performance of a specific CASU during World War II because these units were new and no one knew exactly what to expect. In a larger context, the CASUs were part of shore-based, naval aviation located in the South Pacific, which is to say, this branch of the service was young, inexperienced, and very untested. The Navy simply had not worked out a plan for the administration of its vast air arm ashore. And there was a major missing ingredient, the Navy lacked a consistent written doctrine for shore-based naval air power that was understood across the fleet.

Fortunately, for the United States Navy and for the progress of our war against Japan, there is little evidence that confusion in CASU administration hampered the war effort. Most agree that the Navy was venturing into untested territory by attempting to use islands as supporting units for shore-based and carrier-based naval aviation. No one really knew how it would all work out. At any rate, the islands were used brilliantly and with excellent results. In almost every instance, the island airfields achieved the purpose for which they were captured or built, and they did so amidst the welter of unceasing administrative confusion.

Administrative chaos also existed between the ACORNs and CASUs. According to the plan the ACORN had all the tools and parts while the CASU supplied the trained personnel. Mostly they were physically together and mutually supporting partners. However, sometimes they were on different islands, and occasionally, though only yards apart, the ACORN's tool check-out shack and supply center were closed. The ACORN - CASU team would receive conflicting orders and the chain of command was not always clear.

In their December 1945 War History COMAIRSOPAC reviewed the challenges the CASUs confronted.

The Basic CASU Problems [Just as CASU-11 arrived at Guadalcanal.]

1. An ineffective system for aviation supply operated throughout the Navy until mid-summer 1943 when significant changes were accomplished. However, the benefits of these improvements were not to be felt out in the South Pacific until early 1944.

2. General logistics were poor. In aviation, where many parts are delicate and critical to aircraft performance, poor loading of ships, mis-routing, double and triple handling aggravated an already troubled situation. Proper 'combat loading' techniques, new packaging, and better routing procedures visibly improved the system by summer 1943.

3. Delivery of aircraft was sometimes brilliantly handled and sometimes quite capricious or ineffective. Up until late 1943 or early 1944 it was safe to say that the majority of new types of aircraft reaching the South Pacific arrived at forward bases about six months ahead of supplies and spares to keep them operating. Rarely were instructions for even elementary overhaul available.

4. Few of the early Carrier Aircraft Service Units (CASU's) were qualified to perform the functions for which they had been commissioned. This meant that carrier aircraft which operated with great brilliance aboard their carriers found inadequate service units ashore. Certain CASU's were so incompetent that drastic means had to be initiated to build them up to even the standards of a good automobile garage in a small town. At least five CASU's appeared in the area to maintain and overhaul delicate aircraft like F4U's without even a general knowledge of those planes. Mechanics were asked to make important changes in combat conditions on airplanes they had never seen and which they had no pointed instructions or tables of tolerances.

5. Shore-based aviation in general was under-staffed with trained men, especially aviation personnel, and over-staffed with enthusiastic but unspecialized officers and ordinary seamen.

6. Shore-based aviation was repeatedly assigned to an island in which the island organization (presumably an ACORN, but occasionally a CUB or a LION) was supposed to provide basic facilities such as camp sites, rolling stock, common medical facilities. Frequently the ACORN was actually so busy in the routine procedure of island life that newly arrived aviation

activities had to forage for themselves, when by training, equipment and instruction they were totally unprepared to do so.

7. Most important of all, there was no clear-cut understanding – and there would not be any understanding for more than two and a half years – as to what the relationship should be between various air activities and: (1) other air activities, (2) other naval activities, or (3) activities of other branches of service.[2]

The chain of command over CASU(F)-11 shifted almost daily among on and off island commands. New commands were inserted into the administrative structure with unclear functions, and command names were changed.

However,

On the other [positive] side of the picture.

1. American naval aviation lost no battles in the South Pacific. Even the draw at Coral Sea could be interpreted as a qualified victory.

2. Relationships between all elements of command appear to have been cordial and much more than passably satisfactory.

3. Relationships with the Army were maintained without serious irritation.

4. For each CASU that came out unequipped and unprepared to do its job there was another that went on to some lately occupied island and performed absolutely unbelievable tasks of maintenance and fabrication of jury rigs of one kind or another.[3]

During the period from June 1943 until July 1944 a considerable amount of correspondence and some progress was made to rectify the Navy aviation structure in the South Pacific. However, it was the visit of Artemus L. Gates, Assistant Secretary of the Navy for Air, during the period 16 – 28 May 1943 that caused significant refocusing upon this overall aviation management problem. He stated in his cover letter attached to his staff visit notes,

"Confusion in administrative and organizational matters bearing on aviation supply and distribution in the areas visited has convinced me of an existing lack of appreciation within the operating forces of the magnitude of the logistic problems facing us. Definition of responsibility and grant of authority to accompany it seem to be urgently needed."[4]

One result of this aviation supply problem was observed in CASU-11 on Guadalcanal by the Gates' inspection team, "aviation mechanics (CASU-11 mechanics) roaming with hammers through vast rows of unsheltered boxes of aviation supplies on Guadalcanal, knocking tops off here and there, searching for what they needed. And then there was the 'misappropriation of other people's equipment, materials, and spare parts as a general procedure.'"[5]

The actual misappropriation of other people's equipment by harassed mechanics was not in itself reprehensible. Many important squadron aircraft were kept aloft because some chief had initiative enough to "steal the pants" off some rear echelon area organization. He did so with the full approval of his commanding officer who knew that this was necessary to keep the squadron's airplanes flying. The regrettable aspect of the situation was that sometimes the only way supplies could be obtained was by such misappropriation.

Another management issue was the island's organization might differ significantly from another island's organization. For example, the island commanding officer might be Army, Navy, Marine Corps, or from New Zealand and subject to continuing change. The senior air officer might be Army, Navy, or Marine Corps. Under him, three different air fields might be respectively Army, Navy, and Marine Corps. On an air field, there might be a CASU, a PATSU, a Field overhaul unit from an overhaul activity miles away, maintenance crews from a Marine Air Group, a New Zealand Service Unit, a Maintenance detachment from a V-J Squadron whose headquarters might be 800 miles away, and Army maintenance crews. The Naval Air Center might or might not be the senior control center for ground aviation activities. The commanding officer for the island might or might not have his own air force, and if he had such a force it might be either Army or Navy. Add to the above complexity, four separate supply systems, Army, Navy, Marine, and New Zealand, as well as the possibility of senior aviation officers appearing on the scene at any time to assume varying degrees of command, and the result is a system that is guaranteed to create confusion.

For instance, while CASU-11 was assigned to Guadalcanal, Lt. Cmdr. Schlossbach was commanding officer of Henderson Field (Bomber 1), Lunga Field (Fighter 1), and CASU-11. Maj. Taylor was commanding officer of Kukum Field (Fighter 2) and the 38th Service Squadron. And Lt. Col. Tidwell was the commanding officer of Carney Field (Bomber 2), Koli Field (Bomber 3) and the 29th Army Air Service Group.

Above them in the chain of command and all residing on Guadalcanal was Commander Naval Base Guadalcanal, Commander Naval Bases Forward Areas, Commanding General Forward Areas and Guadalcanal, Commanding General Island Air Command, Commander Forward Air Wing and Air Guadalcanal, Commander Air South Pacific and Commander South Pacific - a guaranteed unwieldy abundance of leadership.

Correspondence generated on the problems associated with the administration of the CASUs in the western Pacific were detailed within the History of Commander Aircraft South Pacific and are summarized below;

The CASU was designed as a highly mobile, streamlined, shore-based unit serving primarily as an aviation garage for aircraft assigned to carriers.

In actuality only 50% of CASU time was spent on airplanes.

A good CASU had to service almost every type of military aircraft.

CASUs rarely moved once they got to their island airfield, however, management always treated the CASU as something that might have to move quickly to another site.

CASU doctrine stated manning should be about 600 men and 30 officers, trained in the states, shipped to an advanced base with a minimum of gear, and use the equipment from a designated ACORN. ACORN and CASU should arrive in the field with all gear and repair parts prior to the arrival of operational [aircraft] squadrons.

A review of the available formal and informal reports on CASU performance are of one or two categories, praised or damned. During 1943 most CASUs were in the damned category, although certain CASUs were selected for favorable and even warmly thankful comments.

All CASUs suffered from administrative indecision on the part of the Navy since all land-based naval aviation was new.

Relationship between CASU and ACORN, while not entirely satisfactory, was judged not to be a serious hindrance to operations.

CASU personnel arrived overseas poorly trained, the schools did not know what to provide and personnel assigned as instructors were uncooperative and suspicious that they might be pushed out to combat if they trained students too well.[6]

Even with proper training the CASU's would still have inadequately trained men, training was on selected aircraft, usually older models, while new aircraft were being shipped overseas.

The ACORN - CASU linkage was doctrinally flawed because every ACORN had a Seabee component to build an airport or repair what was already there. If there was already an available airfield then the Seabee group was not required, if no airfield was in existence then the CASU had to wait until the airfield was constructed.

On 29 September 1943, six months after CASU-11 arrived, Adm Hardison, ComFairSouth reported to his boss at ComAirPac:

"The first CASU's sent to the South Pacific, CASU-3 and CASU-9, were poorly equipped. Although not yet completely equipped in accordance with the latest allowance lists, these units were well trained and have given excellent service. CASU's 10, 11, and 12 were better equipped and had ample time to set up their facilities and train their personnel in well-established areas before receiving any heavy workloads. As a result, they have also given excellent service. CASU's 8 and 14 were by far the best equipped service units to arrive in this area. The performance of CASU-8 in a more forward area has been only fairly satisfactory. As indicated in other correspondence CASU-14 was totally unprepared for its job."[7] (CASU-14 was eventually relieved of duty on Munda Field. After further investigation the primary reason CASU-14 failed was its total lack of training on Marine Corsairs.)

The management skill of the Commanding Officers significantly impacted the performance of the CASU. CASUs were among the most difficult aviation units to administer. Work was always at either an ebb or a flood. The energies of personnel were sometimes sadly wasted in one CASU, whereas another had more work than it could do. Waiting or busily working while always pondering when the next air raid might occur sapped the normal energies of CASU personnel. It took a good officer to command a CASU and hold it to a high level of performance.

Word of these issues began to cycle through the gossip network of the ACORNs and CASUs beginning their training and preparations for deployment overseas. Some units decided to take it upon themselves to solve part of this issue. A later CASU, after CASU-11, had a fuel tanker truck issued to them to take overseas. They proceeded to borrow, lift, abscond, and steal everything they judged they might need from CASU-5 in San Diego and stuff it into the empty fuel tank.

During 1943 - 1944, CASU-11 had a monthly average of approximately 30 aircraft at Henderson Field under their management for repairs and maintenance. Meanwhile, during the period September 1944 to March 1945 when assigned to NAS Thermal, California CASU-11 averaged about 20 aircraft per month. Finally, from April 1945 through late 1946 on Okinawa, aircraft numbers averaged about 10 per month. These weekly aircraft location reports only included the number of airplanes the CASU Commanding Officer had taken charge of during the report period. These reports did not include the often overwhelming numbers of aircraft that showed up unannounced for repairs, fuel and ordnance.

There are 121 Location of U.S. Naval Aircraft reports held at the Naval History and Heritage Command for the period 6 Jan 1943 through 7 Sep 1945. These reports show CASU-11 reported only 1,069 aircraft under their control over these 32 months. It is obvious that there is a significant distinction between aircraft under the control and management of a CASU for this report's purpose and the aircraft that the CASU had drop in for maintenance, rearming and repair. There are just too many examples of land-based and carrier-based squadrons sending letters of thanks for the services CASU-11 and other CASUs performed on their aircraft while the squadron still included these aircraft as under their own reporting responsibility. Additionally, CASUs with online War Diaries almost always reported the number of aircraft they serviced per month greatly exceeded the number they had under their charge requiring bi-weekly inventory reports, so ownership and maintenance numbers are definitely two different numbers.

Appendix I - Chronology

Chronology for CARRIER AIRCRAFT SERVICE UNIT ELEVEN (CASU-11) and COMBAT AIRCRAFT SERVICE UNIT (FORWARD) ELEVEN (CASU(F)-11) during period 22 January 1943 to 1 November 1946.

22 Jan 1943 - CASU-11 commissioned at Naval Air Station San Diego, California. Lt. Cmdr. Isaac Schlossbach assigned as the first Commanding Officer CASU-11.

9 Feb 1943 - SS President Polk received approximately 25 Officers and 550 Enlisted for transport to Guadalcanal, Solomon Islands.

16 Feb 1943 - SS President Polk crossed the Equator while heading west.

23 Feb 1943 - SS President Polk crossed the International Dateline (180 degrees longitude.).

25 Feb 1943 - SS President Polk arrived Espiritu Santo, New Hebrides.

15 Mar 1943 - SS President Polk departed Espiritu Santo for Guadalcanal.

18 Mar 1943 - SS President Polk arrived Guadalcanal.

24 Mar 1943 - CASU-11 dug foxholes at their assigned camp site, number 53, approximately one mile from Henderson Field.

Apr - May 1943 - Defense of Henderson Field and Guadalcanal.

Jun - Sep 1943 - Pause until ready to attack Rabaul.

Jul 1943 - CASU-11 qualified pilots and crew flying combat missions over Munda Field, New Georgia with Navy Squadron VC-28.

Oct 1943 - Mar 1944 - Takedown of Rabaul.

17 Apr 1944 - Lt. Cmdr. Carleton Pike became second Commanding Officer CASU-11, Guadalcanal.

28 Jun 1944 - USS Rochambeau anchors off Lunga Point to pick up first section of CASU-11 to go home.

30 Jun 1944 – USS Rochambeau underway for Espiritu Santo.

2 Jul 1944 – USS Rochambeau arrived Espiritu Santo.

4 Jul 1944 – USS Rochambeau underway for San Francisco, California.

5 Jul 1944 - Lt. Cmdr. Reinhart Vogt became third Commanding Officer CASU-11, Guadalcanal.

13 Jul 1944 – USS Rochambeau passes near typhoon, observes winds of 60 knots.

20 Jul 1944 – USS Rochambeau moored berth 7, pier 31, San Francisco California.

22 Jul 1944 - Lt. Cmdr. Philip Allen Jr became the fourth Commanding Officer CASU-11.

Early Aug 1944 - SS Monterey embarks more personnel of CASU-11 for transport to West Coast. Destination unknown, suspect San Francisco, arrival about 23 Aug 1944. Lt. Cmdr. Vogt moved to CASU 41, Guadalcanal.

29 Aug 1944 - After dispersal of most of CASU-11 personnel to their next duty station or discharge to home, the remainder are transported to ACORN Assembly and Training Detachment, Port Hueneme, California. CASU 11 reforming for another assignment.

25 Sep 1944 - Lt. R. B. Torian is the "Acting Commanding Officer," and fifth Commanding Officer for CASU-11.

25 Oct 1944 - All CASUs outside continental U.S. are renamed Combat Aircraft Service Units (Forward). CASU-11 became CASU(F)-11.

7 Dec 1944 - CASU(F)-11 arrived at Naval Air Facility (NAF) Thermal, California. CASU is receiving new personnel and preparing to return overseas to Okinawa. Unnamed Commanding Officer of CASU-11 committed suicide.

23 Dec 1944 - Lt. Cmdr. E. F. Zimmerman assumed command as sixth Commanding Officer of CASU(F)-11.

Early Jan 1945 - CASU(F)-11 returned to Port Hueneme, California.

12 Feb 1945 - CASU(F)-11 Advanced Echelon (7 Officers including Lt. Cmdr. Zimmerman and 4 Enlisted) loaded onto USS Cepheus (AKA – 18) in Honolulu Harbor. Unknown how CASU(F)-11 Advanced Echelon travelled to Hawaii and where remainder of CASU(F)-11 was located (41 Officers, 844 Enlisted). Lt. Cmdr. Zimmerman remained the Commanding Officer of CASU(F)-11.

26 Feb 1945 – USS Cepheus underway for Guadalcanal.

1 Mar 1945 - Navy Branch Number 13841 established for CASU(F)-11. This enabled mail to follow CASU(F)-11 to Okinawa.

8 Mar 1945 – USS Cepheus arrived Guadalcanal.

15 Mar 1945 – USS Cepheus departed for Ulithi, Carolinas to join task force.

21 Mar 1945 – USS Cepheus arrived Ulithi.

27 Mar 1945 – USS Cepheus departed Ulithi for Okinawa as part of Task Unit 53.2.1 of Transport Group Baker in the Northern Attack Force.

1 Apr - 21 Jun 1945 - Battle for Okinawa, Japan ended after 82 horrific days.

3 April 1945 - CASU(F)-11 Advanced Echelon landed from the USS Cepheus and set-up at Yontan Airfield, Okinawa, Japan.

30 April 1945 - CASU(F)-44 provided three officers and 43 enlisted to assist CASU(F)-11 Advanced Echelon at Yontan Airfield.

12 Apr 1945 - President Franklin D. Roosevelt died at Warm Springs, Georgia.

17 May 1945 - Approximate date Rear Echelon of CASU(F)-11 finally arrived Okinawa with 41 officers and 844 enlisted.

8 Jun 1945 - CASU(F)-44 augmentees departed Yontan Airfield and return to Tinian.

25 Jul 1945 - CASU(F)-11 manning verified as 48 officers and 848 enlisted on this date.

9 Aug 1945 - Carrier Aircraft Pool Detachment CASU (F)-11 with 182 men assigned organized at NOB Okinawa, Navy 3256, Awase Airfield. Cmdr. Tollack was the commanding officer.

2 Sep 1945 - World War II over, peace treaty signed onboard USS Missouri, Tokyo Bay, Japan.

7 Sep 1945 - CASU(F)-11 moved from Yontan Airfield to Yonabaru Field. Yontan Airfield became an Army bomber base, Yonabaru became a Navy air field.

16 Sep 1945 - Typhoon Ida passed about ten miles east of Yonabaru with winds up to 95 miles per hour.

9 Oct 1945 - Typhoon Louise passed about 15 miles east of Yonabaru with winds up to 120 miles per hour.

13 Nov 1945 - Cmdr. Hugh L. Tollack now seventh Commanding Officer of CASU(F)-11.

4 Mar 1946 - Cmdr. R Charles E. Hays assumed command as eighth Commanding Officer of CASU(F)-11.

22 May 1946 - Cmdr. Frederic L. Faulkner assumed command as the ninth Commanding Officer of CASU(F)-11.

1 Oct 1946 - COMFAIRWESTPAC directs decommission of CASU(F)-11 to occur on or about 25 Oct 1946.

15 Oct 1946 - Navy Branch Number 13841 disestablished. Mail ceased being sent to Okinawa for delivery to CASU(F)-11. This inferred that CASU(F)-11 had ended their presence at Yonabaru Airfield.

1 Nov 1946 - Orders executed detaching Cmdr. Faulkner from CASU(F)-11 and directing him to Fleet Aircraft Service Squadron (FASRON) 122 at Yonabaru Airfield for duty as Commanding Officer. CASU(F)-11 was no more, all personnel went to FASRON-122, other duty stations, or home.

Acknowledgements

I'd never considered writing a book. And here I am, over five years into my effort, having done just that. My brother, Tom, first planted the seed that started the research that would lead to this book. His interest in our late father's uniform and the medals he earned during his 20-year career in the Navy made us realize that we knew very little about Dad's wartime experiences. Our dad shared very little about his war service with my brother, sister or I. Even Mom who he met after World War II knew very little. Thank you Tom for nudging me to begin this very rewarding endeavor.

I will be forever grateful for Melissa Hays, who found me through my blog. Her faith and trust in me when she shared her father's CASU-11 Guadalcanal diary was overwhelming. Even when the research materials seemed nonexistent her friendship, continuing interest and encouragement kept me in the game.

To Chief Aviation Metalsmith Harry Hays: This book could not have been written without your personal dedication to writing a few words each day in your diary. The daily glimpses into CASU-11 life on Guadalcanal that you penciled onto the diaries yellowed line pages have been the backbone of this book. The words "thank you" are totally inadequate.

Lynn O'Flynn and Jann O'Flynn Sher are two of my heroes. They also found me through my CASU blog, invited me to their home, and introduced me to their father, Patrick O'Flynn. Their support was wonderful and unlimited. We spent two days together listening to and recording as Patrick told us about his time with CASU-11 on Guadalcanal. Thank you both!

Acknowledgements

Thank you Chief Engineman Patrick O'Flynn. It was one of the greatest honors of my life to meet you and listen to your remembrances of CASU-11 and Guadalcanal. We spent Saturday, 25 July and Sunday, 26 July 2015 together while I recorded almost five hours of your reminisces and answers to questions from Jann, Lynn and I. You have contributed mightily to this book, and it would have been incomplete without your significant contribution.

Steven Rochford, you know this book would not exist without your continuing interest, suggestions and support. Our first interaction was in April 2014. I found you through a small ad you placed in a veteran's magazine. The information on the Internet on CASU-11 was so limited, I cannot express how wonderful it was to find a kindred spirit interested in CASU-11. You have been my continuous wingman, and I stand and salute you for always being there for me, sustaining me, and keeping me moving.

Lynn McAteer Howell, your father's wartime service as CASU-11's Supply Officer does honor to you and your family. Thank you for sharing him with me.

Allison West, you have been an inspiration to me, especially when you shared your desire to fly in a Dauntless Dive Bomber at the Lone Star Flight Museum - I want to go with you.

A very special thanks to Maria Leal. Your love for your Uncle Juan and your interest in knowing what happened to him on Guadalcanal has been an inspiration to me and has reinforced the fact that we must never forget those who sacrificed their all for us and our country.

John Parker and David Wilson - thank you. Your contributions of words and pictures from your fathers has been invaluable. Your personal interest has been very reassuring, especially when I have thought that maybe writing a book was just too much. John, I hope you know you did not have to go all the way to Guadalcanal just for me!

Any work of history requires the cooperation and assistance of many individuals. The following graciously made records available, recommended other avenues of research, provided timely advice and suggestions, and reproduced many copies of source materials and photos.

Terry L. Potter - Archivist, Independence Seaport Museum, Philadelphia, Pennsylvania.

Sara Stopper - Administrative Assistant to Ms. Potter.

Dennis Rochford - Nephew of John McAteer, and President of Maritime Exchange, Philadelphia, Pennsylvania.

Jessica Doney - Photo Technician, CVS Drugstore, Springfield, Pennsylvania. Jessica kindly reproduced prints of all the photographs left by Lt. McAteer.

Volunteers at Independence Seaport Museum, also known as "The Friday Guys"

Dave Kavanagh - Research partner of Steve Rochford and Sun Ship Historical Society.

David Boone - Research partner of Steve Rochford.

John Costello - Research partner of Steve Rochford.

Mike Kunz - Your stories shared with Cindy, Dale, Sally and I that special evening in Phoenix will stick with me forever. Your service with CASU-49 was honorable and in accordance with the highest tenets of the United States Navy. Well done, shipmate, and thank you.

Gina Bardi - Reference Librarian, San Francisco Maritime National Historical Park Research Center. Your professional assistance in locating and interpreting the deck logs of the SS President Polk was very much appreciated.

William Greene and Aaron Seltzer - Archival Staff, National Archives at San Francisco. Your knowledge of the archival holdings covering U.S. Navy bases on foreign territory in the Pacific was right on the mark. Every question was swiftly answered by William and Aaron who placed just the right file box in front of me.

Peter Flahavin - From Australia. You have been a critical part of my support system while researching this book. Your collection of current and World War II era pictures have been extremely valuable to me as I gained an understanding of Guadalcanal. Your responsiveness to my intermittent emails has been much appreciated.

Mark Goodman - Thank you for introducing me to John Barnard Goodman Jr. and his assignment at CASU (F)-11 in Okinawa in 1945, and sharing the existence of the Special Weapons Ordnance Device "Bat."

Donn [Declined to provide last name.] - Thanks for the access to your "Remembering Okinawa" web site and permission to use some of your pictures. By providing me with the military mail service start and end dates for CASU (F)-11 on Okinawa, you helped me correctly determine the length of time CASU (F)-11 was actually on the island.

Greg Bingham - Your timely contact with pictures of CASU(F)-11 on Okinawa greatly enhanced the telling of the story about the 24 May 1945 twelve aircraft commando raid on Yontan Airfield. Thank you.

Andrea "Andy" Sikkink of AS You Wish Arts volunteered to create the front cover for this book. Your picture ideas and color selections were outstanding. You asked all the right questions and your decision to use colors from the Asiatic-Pacific Campaign and World War II Victory medals was perfect. You have my sincere gratitude. And Anna, thank you too for your timely, last minute assistance.

Finally, to my family. It is an incredible person that can take almost 400 pages of miscellaneous facts and photos on a Navy aircraft repair unit during World War Two and turn them into a coherent story. It requires patience and devotion combined with dedicated reading and editing. Every reader can be assured every word was critiqued by an outstanding editor, my bride of forty nine years, Sally Schoppert Little. I love you.

Acknowledgements

Jessica, who after listening to story after story, encouraged me to write a blog which rapidly brought me into contact with numerous people interested in CASU-11. She was my editor, who provided ongoing advice for generating a book with perfect format, grammar, sentence structure and punctuation. Without her participation there would not have been a completed book; Jessica, you are amazingly talented. I love you.

Rachel, thank you for your unceasing, positive encouragement and the new granddaughter Elise Isabella. I love you.

The three of you have been lifelong blessings beyond measure.

Acknowledgements

Notes

Notes for entire book

1. Hays, H.A. (1944). *Diary of Harry Hays CASU-11, February 1943 to August 1944* provided by his daughter Melissa Hays.

2. O'Flynn, P. G. (2015). *Recorded Interview.* Transcript of recorded interview conducted 25 and 26 July 2015 at his residence in Tampa Bay, FL.

3. Little, R. H. (1982). *Family cassette tapes.* Transcripts created from 1980-1982 recordings held by his son, William H. Little.

Notes to Introduction

1. Little, R.H. (1962). *Navy Service Record for Robert H. Little, Serial Number 602-09-40.* Copy held by William H. Little.

Notes to Part One. San Diego, CA

1. Commanding Officer CASU Five. (1945). Letter serial CASU5/A9-8 dated 31 January 1945. Retrieved from https://www.fold3.com/image/251/302025397 Page 14.

2. Commanding Officer CASU Five. (1945). Page 14

3. Nash, G.D. (1990). *The American West Transformed: The Impact of the Second World War.* Lincoln and London, NE: University of Nebraska Press. Page 59.

Notes to Part Two. Espiritu Santo, New Hebrides

1. SS President Polk (AP-103). (1943). *Deck Logs on the February – March 1943 Voyage.* Archived at the San Francisco Maritime National Park Research Center, Fort Mason, Landmark Building E, 2 Marina Blvd, San Francisco, CA 94109.

2. Naval Supply Depot San Diego. (1943). *War Diary Feb 1943.* Retrieved from https://www.fold3.com/image/251/268514762.

3. Phillips, Bum. (2010). *Bum Phillips: Coach, Cowboy, Christian.* Brenham, TX: Lucid Books. Pages 44-45.

4. Marsh, D.Z. (1943). *The Becoming.* Retrieved from the "original Marine Raider website" http://www.usmcraiders.org/the-becoming/.

5. Rigsby, J.M. (2006). *Collier County WWII Capture Living History Project.* Interview retrieved from https://www.colliercountyfl.gov/your-government/divisions-a-e/communication-customer-relations-division/wwii-capture-living-history/john-m-rigsby.

6. Rigsby (2006).

7. Commanding Officer Armed Guard Unit. (1942). *SS President Coolidge Letter December 17, 1942.* Retrieved from http://www.armed-guard.com/cool.html. Pages 2-3.

8. The Bluejackets' Manual. (1943). United States Naval Institute, Annapolis, MD, 1943. Eleventh Edition. Page 415.

9. Commander Air Center, Navy 140. (1945). Letter CAC/A12 Serial 580 dated 13 June 1945. Retrieved from https://www.fold3.com/image/251/302003663. Army enclosure. Page 20.

10. Commander Air Center, Navy 140. (1945). Page 23. Mr. Matt Well's enclosure.

11. Commander Air Center, Navy 140. (1945). Page 20. Army enclosure.

12. Commander Air Center, Navy 140. (1945). Page 23. Mr. Matt Well's enclosure.

13. Staff Historical Officer, South Pacific Area and Force. (1945). *Notes on Visit to South Pacific Area, 16-28 May, 1943*. Retrieved from https://www.fold3.com/image/1/302031591. Page 33.

14. Marsh (1943). *Mudding Through*. Retrieved from the "original Marine Raider website" http://www.usmcraiders.org/the-becoming/.

15. Merillat, H.C. (1982). *Guadalcanal Remembered*. New York, NY: Dodd, Mead & Company. Pages 32-33.

16. Isenberg, F.J. (2016). *Navy Service Record for Ferdinand J. Isenberg, Serial Number 644-04-41*. Read at National Military Service Record Center, St Louis MO.

17. Commander in Chief, U.S. Pacific Fleet and Pacific Ocean Areas. (1943). *Operations in Pacific Ocean Areas*. Monthly Reports, February 1943 through February 1946.

18. Fussell, P. (1989). *WARTIME Understanding and Behavior in the Second World War*. New York Oxford: Oxford University Press. Page 20.

Notes to Part Three. Guadalcanal, Solomon Islands

1. Rigsby, J.M. (2006).

2. Prados, John. (2003). *Torpedo Junction*. HISTORYNET Blog 3/4/2003. Retrieved from
https://www.historynet.com/torpedo-junction.htm.

Notes

3. Prados. (2003).

4. Prados. (2003).

5. USS Ellet (DD 398). (1943). *War Diary March 1943*. Retrieved from https://www.fold3.com/image/251/268975441.

6. USS Breese (DM 18). (1943). *War Diary March 1943*. Retrieved from https://www.fold3.com/image/251/268754391 and 93 and 95.

7. USS Morris (DD 417). (1943). *War Diary March 1943*. Retrieved from https://www.fold3.com/image/251/268925162.

8. (Keese & Sidwell, 1991). *United States History for Christian School* (2nd Ed.). Greenville, SC: Bob Jones University Press. Page 528.

9. Rentz, J.M. (1952). *Marines in the Central Solomons.* USMC Historical Monograph Chapter 1, Introduction. Page 1.

10. Prados, J. (2012). *ISLANDS OF DESTINY: The Solomons Campaign and the Eclipse of the Rising Sun*. London, England: Penguin Books. Page 16.

11. Building the Navy's Bases in World War II*: History of the Bureau of Yards and Docks and the Civil Engineer Corps, 1940-1946*: Washington, DC: United States Government Printing Office   Retrieved from http://www.ibiblio.org/hyperwar/USN/Building_Bases/index.html#contents2. Chapter XXV. Pages 241 - 244.

12. 6th Naval Construction Battalion Historical Information. (1945). U.S. Navy Seabee Museum, 3201 N Ventura Road, Port Hueneme, CA 93043. https://www.history.navy.mil/content/dam/museums/Seabee/UnitListPages/NCB/006%20NCB.pdf. Page 14.

13. Pistol Pete - artillery (150mm Howitzer) landed by Japanese 4th Heavy Field Artillery Regiment to prevent Allied forces from establishing and operating Henderson Field. Seriously damaged Henderson Field mid

October 1942. Finally destroyed by Lt. Col Evans Carlson and his Marine Raiders on 30 November 1942.

14. 46th Naval Construction Battalion Historical Information. (1945). U.S. Navy Seabee Museum, 3201 N Ventura Road, Port Hueneme, CA 93043. https://www.history.navy.mil/content/dam/museums/ Seabee/ UnitListPages/NCB/046%20NCB.pdf. Page 30.

15. 61st Naval Construction Battalion Historical Information (1945). U.S. Navy Seabee Museum, 3201 N Ventura Road, Port Hueneme, CA 93043. https://www.history.navy.mil/content/dam/museums/ Seabee/UnitListPages/NCB/061%20NCB.pdf. Pages 1 to 19.

16. 18th Naval Construction Battalion Historical Information. (1945). U.S. Navy Seabee Museum, 3201 N Ventura Road, Port Hueneme, CA 93043. https://www.history.navy.mil/content/dam/museums/ Seabee/UnitListPages/NCB/018%20NCB.pdf. Pages 33 and 34.

17. 14th Naval Construction Battalion Historical Information. (1946). U.S. Navy Seabee Museum, 3201 N Ventura Road, Port Hueneme, CA 93043., https://www.history.navy.mil/content/dam/museums/ Seabee/UnitListPages/NCB/014%20NCB.pdf. Page 19.

18. National Archives at San Francisco, CA, 1000 Commodore Drive, San Bruno, CA, 94066-2350, (https://www.archives.gov/san-francisco). Loose handwritten undated document detailing Field Commander assignments for Guadalcanal found within archival box filled with administrative files for 1943/1944.

19. Manual of Advanced Base Development and Maintenance. (1945). OpNav 30-11-A1 of July 1943 (Revised Apr 1945). Retrieved from https://archive.org/details/manualofadvanced00unit Pages 4-5 and 106.

20. Manual of Advanced Base Development and Maintenance. (1945). Page 106.

21. Naval History and Heritage Command. (1944). *Naval Aviation News*. September 1, 1944 Restricted. Retrieved from

https://www.history.navy.mil/content/dam/nhhc/research/histories/naval-aviation/Naval%20Aviation%20News/1940/pdf/1sep44.pdf, Page 16.

22. Bergerud, E.M. (2000). *Fire in the Sky, The Air War in the South Pacific*, Boulder, CO: Westview Press. Page 6.

23. Bergerud. (2000). Page 23.

24. Naval History and Heritage Command. (1944) Page 15.

25. Naval History and Heritage Command. (1944). Page 18.

26. U.S. Naval Advanced Base Unit. (1943). *ACORN (RED) One, War Diary*. Dated 12 October 1943. Retrieved from https://www.fold3.com/image/1/267921013.

27. U.S. Naval Advanced Base Unit. (1943). *ACORN (RED) One, War Diary*. Dated 23 March 1943. Retrieved from https://www.fold3.com/image/1/268686517.

28. Merillat. (1982). Page 1.

29. U.S. Naval Advanced Base Unit. (1943). *ACORN (RED) One, War Diary*. Dated 9 June 1943. Retrieved from https://www.fold3.com/image/1/269442858.

30. Leckie, R. (1957). *Helmet for my Pillow*. New York: Simon & Schuster. ISBN 0-7394-1404-6. Pages 91-92.

31. Bergerud. (2000). Pages 79-80.

32. Astor, G. (2005). *SEMPER FI IN THE SKY The Marine Air Battles of World War II*. New York: Presidio Press. ISBN 0-7394-6393-4. Page 83.

33. Foster, J. M. (1961). *Hell in the Heavens*. New York, NY: G.P. Putnam's Sons. Pages 17-18.

34. Astor. (2005). Pages 199-200.

35. Wukovits, J. F. (2011). *Black Sheep: the life of Pappy Boyington.* Annapolis, MD: Naval Institute Press. ISBN 978-1-59114-980-4. Page 111.

36. Bergerud. (2000). Page 101.

37. Kreis, J. F. *(1988). Air Warfare and Air Base Defense* 1914 – 1973. Washington, D.C: U.S. Government Printing Office. Page 231.

38. Kreis. (1988). Page 231.

39. Kreis. (1988). Page 229.

40. Morison, S. E. (1950). *Breaking the Bismarcks Barrier.* Champaign, IL: University of Illinois Press. ISBN 0252069978. Page 102.

41. Taylor, T. (1954). *The Magnificent Mitscher.* New York, NY: W.W. Norton & Co. ISBN 1557508003. Pages 147-148.

42. Bennett, J. A. (2009). *Natives and Exotics World War II and Environment in the Southern Pacific.* Honolulu, HI: University of Hawaii Press. Page 97 and 99.

43. Merillat. (1982). Page 131.

44. Moore, S. L. (2014). *Pacific Payback.* New York, NY: Penguin Group. ISBN 978-0-451-46552-8. Page 53.

45. Tillman, B. (1976). *The Dauntless Dive Bomber of World War II.* Annapolis, MD: United States Naval Institute. ISBN 0-87021-569-8. Page 12.

46. Bergerud. (2000). Pages 295-296.

47. Joy, R. J. T. (1999). *Malaria in American Troops in the South and Southwest Pacific in World War II.* Journal of Medical History. Vol 43: 192 - 207. Page 199.

48. Leckie. (1957). Page 162.

49. Mersky, P. B. (1997). *U.S. Marine Corps Aviation 1912 to the Present*. Baltimore, MD: The Nautical & Aviation Publishing Company of America, 1997. Pages 89 - 90.

50. Lane, K. L. (2004). *GuadalcanalMarine*. Jackson, MS: University Press of Mississippi. Page 186.

51. Potter, E. B. (1960). *SEA POWER - A Naval History*. Prentice Hall Inc.: Englewood Cliffs, NJ. Page 711.

52. Morison, (1950). Page 89.

53. Bergerud. (2000). Page xix.

54. Morison. (1950). Page 90.

55. Morison. (1950). Page 90.

56. Potter. (1960). Page 731.

57. Leckie. (1957). Page 109.

58. Leckie. (1957). Page 134.

59. Leckie. (1957). Page 143

60. Morison. (1950). Page 100.

61. McCleary, E. E. (1945). *History of US Naval Advanced Base Guadalcanal 1942-1945*, U.S. Navy Report. Microfiche page 210.

62. Craven, W. F. & Cate, J. L. (1958). *The Army Air Forces in World War II*. Vol IV, *The Pacific: Guadalcanal to Saipan August 1942 to July 1944,* Section II: Target Rabaul, Chapter 7: The Central Solomons. Pages 217-218. https://www.ibiblio.org/hyperwar/AAF/IV/AAF-IV-7.html.

63. Boyington, G. (1958). *Baa Baa Black Sheep*. New York, NY: G.P. Putnam's Sons. Page 178.

64. Bergerud. (2000). Page 153.

65. Torpedo Squadron 28. (1945). *History of Torpedo Squadron Twenty - Eight*. Retrieved from https://www.fold3.com/image/302038275.

66. Taylor, T. (1954). Page 161.

67. Budge, K. G. (2009). *The Pacific War Online Encyclopedia.* Retrieved from http://pwencycl.kgbudge.com/M/e/Medicine.htm.

68. Bennett. (2009). Pages 254-255.

69. Leckie. (1957). Page 86.

70. Lane, K. (2004). *Marine Pioneers: The Unsung Heroes of World War II.* Atglen, PA:Schiffer Publishing, Ltd.

71. Wolfert, I. (1943). *Battle for the Solomons.* Boston, MA: Houghton Mifflin Company. Page 50.

72. Foster. (1961). Page 59.

73. Bennett. (2009). Pages 244-245.

74. Bennett. (2009). Page 257.

75. Bennett. (2009). Page 264.

76. WARFARE HISTORY NETWORK, WWII. (2016). *Eleanor Roosevelt: American Ambassador to the South Pacific.* Retrieved from http://warfarehistorynetwork.com/daily/wwii/eleanor-roosevelt-american-ambassador-to-the-south-pacific/.

77. Youngs, J. W. T. (2013). *Eleanor Roosevelt: A Personal and Public Life. American Realities with Bill Youngs, Prologue: The South Pacific, 1943.* Retrieved from http://www.americanrealities.com/eleanor-roosevelt-south-pacific.html.

78. Gailey, H. A. (1995). *The War In The Pacific.* Novato, CA: Presidio Press. ISBN 0-89141-486-X. Page 224.

79. Cressman, R. J. (1999). *The Official Chronology of the U.S. Navy in World War II.* Washington, DC: Naval Historic Center. Retrieved from https://www.ibiblio.org/hyperwar/USN/USN-Chron.html. See 11 October 1943 entry.

80. Aster. (2005). Page 200.

81. Parker, J. R. (2017). *CASU-11.* Retrieved from http://www.the bestofourlives.com/ of 15 December 2017.

82. Foster (1961). Pages 55-56.

83. Potter. (1960). Page 731.

84. Wilson. D. A. (2015). *Navy Service Record for David A. Wilson, Serial Number 642-10-87.*

85. Butler, J. (2001), *Fire, Smoke and Steel, The Jungle-Fighting 82nd Chemical Mortar Battalion.* Retrieved from https://www.4point2.org/hist-82-p1.htm.

86. Bennett. (2009). Page 206.

87. Fussell. (1989). Page 99.

88. Astor. (2005). Page 145.

89. Fussell. (1989). Page 101.

90. Fussell. (1989). Page 102.

91. Astor. (2005). Page 185.

92. Goodrich, P. (1974). *Ike's travels: The real life story of Navy Commander Isaac Schlossbach, U.S.N. retired, a pioneer submarine*

commander, dive bomber pilot, and explorer ... one trip to the jungles of Central America. Neptune, NJ: Township of Neptune. Portion of this book covering Schlossbach's aircraft flight accident read to author by UCLA Special Collections librarian. Paraphrased version provided from those spoken words.

93. Parker, J. R. (2015). *CASU 11*. Retrieved from http://www.thebestofourlives.com/ of 24 January 2015.

94. Mersky. (1997). Pages 93-94.

95. Navy and Marine Corps Awards Manual, NAVPERS 15,790 (REV. 1953). Page 59. Retrieved from https://www.ibiblio.org/hyperwar/USN/ref/Awards/Awards-IV-16.html

96. Navy and Marine Corps Awards Manual, NAVPERS 15,790 (REV. 1953). Page 9. Retrieved from https://www.ibiblio.org/hyperwar/USN/ref/Awards/Awards-I.html#sec4

97. USS Rochambeau. (1944). *War Diary, Month of July, 1944.* Retrieved from https://www.fold3.com/image/251/279794757. Page 2.

Notes to Part Four. Port Hueneme and Thermal, CA

1. Blazich, F. A. (2014). Historian U.S. Navy Seabee Museum, Seabee Magazine 26 November 2014. Cover Feature *Harbor-Base-Neighbors: When the Navy Came to Port Hueneme, 1942-1945, and Beyond.* Retrieved from http://seabeemagazine.navylive.dodlive.mil/2014/11/26/harbor -base-neighbors-when-the-navy-came-to-port-hueneme-1942-1945-and-beyond/

2. Shettle, M. L. (1944) *NAF Thermal History.* Retrieved from http://www.militarymuseum.org/ThermalAAF.html. Page 1.

3. Shettle. (1944). Page 1

4. Shettle. (1944). Page 1.

5. Combat Aircraft Service Unit (F)-14. (1945). *13 Dec 44 - 2 Sept 45 War Diary*. Retrieved from https://www.fold3.com/image/302101678. Page 2.

Notes Part Five. Okinawa, Japan

1. Appleman, R.E., Burns, J.M., Gugeler, R.A., Stevens, J. (1993). *Okinawa: The Last Battle*. United States Army in World War II, The War in the Pacific. Washington, DC: Center of Military History. Page 36.

2. Potter, E. B. and Nimitz, C. W. (1960). *Sea Power - A Naval History*, Englewood Cliffs, NJ:Prentice-Hall, Inc. Page 828.

3. Yonabaru Naval Air Base. (1945). *Pictorial Souvenir Booklet.* Okinawa: U.S. Navy. Page 4. Retrieved from https://www.rememberingokinawa.com/page/1945_yonabaru_nas.

4. Morison, S. E. (1990). *Victory in the Pacific.* History of United States Naval Operations in World War II, Vol XIV. Boston, MA: Little, Brown and Company. Page 83.

5. Morison. (1990). Page 86.

6. Feifer, G. (1992). *Tennozan: the Battle of Okinawa and the Atomic Bomb*. New York, NY: Houghton, Mifflin Co. Page 133-134.

7. Feifer. (1992). Page 145.

8. 58th Naval Construction Battalion. (1956). *Historical Information.* Retrieved from https://www.history.navy.mil/content/history/museums/ seabee/explore/seabee-unit-histories/ncb.html. Page 42.

9. Witness to War, Combat Stories from World War II. (2001). *Frank Nilson ACORN 29 – Navy, Video recording.* Retrieved from http://www.witnesstowar.org/combat_stories/WWII/1244.

10. Commander Air Pacific Sub Command Forward. (1945). *Air Logistic Plan, No.1-45 Annex "A"*. Retrieved from https://www.fold3.com/image/251/302032739. Pages 3.

11. Commander Air Pacific Sub Command Forward. (1945). Page 4-5.

12. Commanding Officer, Combat Aircraft Service Unit (F) Forty Four. History of CASU (F) 44 - Supplement to. 1 Sep 1945. https://www.fold3.com/image/251/302032790. Page 1.

13. Morison. (1990). Page 224.

14. Morison. (1990). Page 200.

15. Commanding Officer, USS Bellow Wood (CVL-24). (1945). *Action Report 14 March - 28 April*. Retrieved from https://www.fold3.com/image/295838290.

16. Commanding Officer, Patrol Bombing Squadron 21. (1945). *War Diary 1 - 30 April 1945*. Retrieved from https://www.fold3.com/image/296167403.

17. Feifer. (1992). Page 293.

18. Leckie. (1995). Page 126.

19. Commander, Naval Air Bases, Navy No. 3256. (1945). *Action Report, Phase I, OKINAWA Operation*. Retrieved from https://www.fold3.com/image/300791140.

20. Commander Task Group Fifty-Eight Point Four. (1945). *Letter Serial 0211 of 2 May 1945*. Pages 1 and 2. From files Regional Archive Office San Francisco, CA.

21. Commander, Naval Air Bases, Navy No. 3256. (1945). *War Diary of May, 1945*. Retrieved from https://www.fold3.com/image/251/296062783.

22. Morison. (1990). Page 270.

23. Sledge, E.B. (1981). *With The Old Breed at Peleliu and Okinawa.* New York, NY: Oxford University Press. Page 265.

24. Leckie, R. (1995). *Okinawa - The Last Battle of World War II.* New York, New York: Penguin Books. ISBN 978-0-14-017389-5. Page 165.

25. Feifer. (1992). Page 305.

26. Commander, Naval Air Bases, Navy No. 3256. (1945). *Letter Serial F15/NAB3256 of 8 May 1945.* From files Regional Archive Office San Francisco, CA.

27. Commanding Officer, Patrol Bombing Squadron 109. (1945). *Patrol Bombing Squadron ONE HUNDRED NINE - History of.* Retrieved from https://www.fold3.com/image/251/302040934 Page 75.

28. Commanding Officer, Patrol Bombing Squadron 109. (1945). Page 77.

29. Commanding Officer, Patrol Bombing Squadron 109. (1945). Page 76.

30. Feifer. (1992). Page 326.

31. Marine Air Group 31. (1945). *War Diary May 1 to 31, 1945.* Retrieved from https://www.fold3.com/image/251/296078112. Page 4.

32. Marine Air Group 31. (1945). Page 5.

33. Feifer. (1992). Page 390.

34. Leckie. (1995). Page 184.

35. Appleman, R.E., Burns, J.M., Gugeler, R.A., Stevens, J. (1993). *Okinawa: The Last Battle.* United States Army in World War II, The War in the Pacific. Washington, DC: Center of Military History. Page 361.

36. Commander Naval Air Bases, Navy No. 3256. (1945). *War Diary month of May, 1945*. Retrieved from https://www.fold3.com/image/251/296062809 Page 1.

37. Commanding Officer, Combat Aircraft Service Unit (F) Forty Four. (1944). *CASU(F)-44 - History of*. Retrieved from https://www.fold3.com/image/1/302031771 Chronology pages 1 - 2.

38. Commander Naval Air Bases, Navy No. 3256. (1945). Pages 1 and 2.

39. Commanding Officer, Patrol Bombing Squadron 109. (1945). Page 82.

40. Commanding Officer, Patrol Bombing Squadron 123. (1945). *War Diary of Patrol Bombing Squadron 123 for the month of June 1945*. Retrieved from https://www.fold3.com/image/251/302041969. Page 5.

41. Commanding Officer, Patrol Bombing Squadron 124. (1945). *Unit History – Submission of*. Retrieved from https://www.fold3.com/image/251/300855308. Page 4

42. Morison. (1990). Page 233.

43. Commander Task Group 58.4. (1945). *Damaged Aircraft Landing at YONTAN and other fields at OKINAWA*. Letter. From files Regional Archive Office San Francisco, CA.

44. Feifer. (1992). Page 491.

45. Commanding Officer, Patrol Bombing Squadron 109. (1945). *War History 8/3/43 – 10/5/44*. Pages 89-90.

46. Commanding Officer, Patrol Bombing Squadron 118. (1945). *Summary of Operations, Patrol Bombing Squadron ONE HUNDRED EIGHTEEN*. Retrieved from https://www.fold3.com/image/251/300850418.

47. Commander in Chief, U. S. Pacific Fleet and Pacific Ocean Areas. (1946). *Report of Surrender and Occupation of Japan*. Retrieved from https://www.fold3.com/image/251/302077359.

48. 145th Naval Construction Battalion. (1945). *Historical Information*. Retrieved from https://www.history.navy.mil/content/dam/museums/ Seabee/UnitListPages/NCB/145%20NCB.pdf. Pages 100 and 101.

49. Commanding Officer, Patrol Bombing Squadron 109. (1945). *War History 8/3/43 - 10/5/44*. Page 6.

50. Farelly, E. (2016). War History Online. (2018). *Bringing Home The 8 Million Boys After WWII; Operation Magic Carpet*. Retrieved from https://www.warhistoryonline.com/world-war-ii/bringing-home-8-million-boys-.html.

51. Milstein, S. B. (2008). UNIVERSAL SHIP CANCELLATION SOCIETY DATA SHEET #31. *OPERATION MAGIC CARPET*. Retrieved from https://www.uscs.org/wp-content/uploads/2012/04/DS31_Operation - MagicCarpet.pdf.

52. Commander in Chief, U. S. Pacific Fleet and Pacific Ocean Areas. (1946). *Report of Surrender and Occupation of Japan. 11 February 1946. Annex A*. Retrieved from https://www.fold3.com/image/302077826.

53. Commander Naval Air Bases, Navy No. 3256. (1945). *Commendation Letter*. From files Regional Archive Office San Francisco, California.

54. Navy and Marine Corps Awards Manual, NAVPERS 15,790 (REV. 1953). Page 9. Retrieved from https://www.ibiblio.org/hyperwar/USN/ref/Awards/Awards-I.html#sec4.

55. U.S. Navy Patrol Squadrons Webpage. (2000). *A Bit of History*. Retrieved from https://www.vpnavy.com/casu_1940.html. Circa 1946.

56. Naval Aviation News Magazine. October 1946. *BETTER MAINTENANCE REQUIRED BY I.A.P [Integrated Aeronautics Program].* Page 36.

57. Naval Aviation News Magazine. August 1946. *NAVAL AIR HAS MANY CHANGES SINCE V-J [Victory - Japan] DAY.* Page 18.

58. Naval Aviation News Magazine. September 1946. *IAP SUCCESS HINGES ON A&R [Assembly and Repair].* Page 23.

59. Naval Aviation News Magazine. October 1946. Page 36.

60. Navy and Marine Corps Awards Manual, NAVPERS 15,790 (REV. 1953). Page 59. Retrieved from https://www.ibiblio.org/hyperwar/USN/ref/Awards/Awards-IV-16.html.

61. Navy and Marine Corps Awards Manual, NAVPERS 15,790 (REV. 1953). Page 161. Retrieved from https://www.ibiblio.org/hyperwar/USN/ref/Awards/Awards-IV-17.html#sec2-17.

Notes Part Six. Epilogue

1. Laurence, John. *The Cat From Hue.* New York, NY: PublicAffairs, 2002. ISBN 9780786724680. Page 718.

Notes Appendix A. CASU-11 Contributing Members and Group Photos

1. No notes.

Notes Appendix B, Chief Metalsmith Hays joins CASU-9

1. No notes.

Notes Appendix C. Commanding Officers of CASU-11

1. Schlossbach, I. (1984). *Navy Service Record for Isaac Schlossbach, Serial Number 8973.*

2. Pike, C. M. (1967). *Navy Service Record for Carleton M. Pike, Serial Number 26055.*

3. Vogt, R. E. (1981). *Navy Service Record for Reinhart E. Vogt, Serial Number 62360.*

4. Allen, P. (1963). *Navy Service Record for Philip Allen Jr, Serial Number 37335.*

5. Torian, R. B. (1979). *Navy Service Record for Raymond B. Torian, Serial Number 118179.*

6. Unknown Officer lost by suicide.

7. Zimmerman, E. F. (1967). *Navy Service Record for Eugene F. Zimmerman, Serial Number 72029.*

8. Tollack, H. L. (1987). *Navy Service Record for Hugh L. Tollack, Serial Number 170496.*

9. Hays, C. E. (1971). *Navy Service Record for Charles E. Hays, Serial Number 149092.*

10. Faulkner, F. L. (2007). *Navy Service Record for Frederic L. Faulkner, Serial Number 79572.*

Access to the above service records was at the National Archives at St. Louis, MO, 1 Archives Drive, St. Louis, MO 63138, (stl.archives@nara.gov). Visits to view Official Military Personnel Files occurred on various dates from 2014 through 2019.

11. Schlossbach, I. Biography. Retrieved from https://en.wikipedia.org/wiki/Isaac_Schlossbach.

12. Schlossbach. Biography.

13. Schlossbach. Biography.

14. Schlossbach, I. Retrieved from https://www.findagrave.com/memorial/35713163/isaac-schlossbach

15. Maine Historical Society (US). *Lt. Commander Carleton Pike, Lubec, ca.1944*. Retrieved from https://www.mainememory.net/artifact/36847.

16. Combat Aircraft Service Unit (F)-14. (1945). *13 Dec 44 - 2 Sept 45 War Diary*. Retrieved from https://www.fold3.com/image/302101683. Page 4.

17. Eugene F. Zimmerman Papers (WUA00104), 1911-1968 | WUA University Archives. Retrieved from http://archon.wulib.wustl.edu/?p=collections/findingaid&id=175&q =&rootcontentid =302108#id302108

18. Yonabaru Naval Air Base. *Pictorial Souvenir Booklet, Okinawa: U.S. Navy.1945*. Page 4. Retrieved from https://www.rememberingokinawa.com/page/1945_yonabaru_nas.

19. Faulkner, F. L. (2007). *Navy Service Record for Frederic L. Faulkner, Serial Number 79572*. Award letter inserted into service record 3 Oct 1942.

20. Lundstrom, J. B. (2006). *Black Shoe Carrier Admiral - Frank Jack Fletcher at Coral Sea, Midway, and Guadalcanal*. Annapolis, MD: Naval Institute Press. ISBN 9781591144199. Chap 15, first page.

Notes Appendix D. Awards and Commendations

1. Taylor, T. (1954). The Magnificent MITSCHER. Annapolis, MD: United States Naval Institute. Page 161.

2. A review of several dozen service record photographs of CASU-11 men, who served on Guadalcanal, show them wearing their medals. The men have two, three, and even four engagement stars on their Asiatic-Pacific Campaign medal, leaving a mystery of why or how they merited them. Service records only provide certification of the member's authority

to wear the Asiatic-Pacific Campaign medal, rarely does the record have engagement star information.

The Navy Awards Manual provides specific guidance on the dates, locations and military units eligible for Asiatic-Pacific engagement stars. It lists "Consolidation of Solomon Islands between 8 February 1943 and 15 March 1945" as one of the possible selections for an engagement star. It seems that this is the only selection that applies to CASU-11 and means CASU-11 men who served on Guadalcanal should wear one star. However, CASU-11 is not included on the list of eligible units.

This matter was possibly the subject of an investigation by a Navy Ad Hoc Research Group which published the following two volume report in 1951. *Carrier Aircraft Service Units, Fleet Air Wings, Fleet Airship Wings, and Respective Headquarters Squadrons and Supporting Units Ad Hoc Research Group*, "Reports." 2 vols. Individual Personnel, 1951. 225 pp.

This research group, headed by Captain D.F. Smith, undertook extensive investigation of the World War II operations of the indicated units in order to determine eligibility for unit and individual awards. The reports present detailed statistical and narrative information on the experiences of these commands. One volume related to Carrier Aircraft Service Units (CASUs), while the other covered the remaining commands assigned to this study.

This report was included on a list of Navy historical documents claimed to have been moved sometime prior to 1977 from the Naval History Division to the National Archives in College Park Maryland. Contact with the National Archives verified the fact that they do not hold this two volume set. They further shared that they had never received this specific document. Contact with the Naval History Division library also verified they no longer hold a copy of this information. It is now officially missing.

More specifically, while the volume covering Carrier Aircraft Service Units has gone missing, copies of the remaining research are readily available at several libraries. One can only wonder if this research effort possibly resulted in the additional engagement stars most CASU-11 members wore upon their dress uniforms years after World War II.

Notes Appendix E. Ships That Transported CASU-11

<u>SS President Polk</u>

1. NavSource Online. (Polk). Retrieved from http://www.navsource.org/archives/09/22/22103.htm.

2. California Digital Library. *Guide to the American President Lines Records, 1871-1995.* Retrieved from https://oac.cdlib.org/findaid/ark:/13030/tf4j49n761/.

<u>USS Munargo</u>

3. NavSource Online. (Munargo). Retrieved from http://www.navsource.org/archives/09/22/22020.htm.

4. Circuit Court of Appeal, Second Circuit. Retrieved from https://www.leagle.com/decision/193754490f2d4541408.xml.

5. NavSource Online. (Munargo).

<u>USS General Harry Taylor</u>

6. NavSource Online. (Taylor). Retrieved from http://www.navsource.org/archives/09/22/22145.htm.

7. Naval History and Heritage Command. Retrieved from https://www.history.navy.mil/research/histories/ship-histories/danfs/g/general-harry-taylor.html.

<u>USS Rochambeau</u>

8. NavSource Online. (Rochambeau). Retrieved from http://www.navsource.org/archives/09/22/22063.htm

9. Charles, R.W. (1947). *Troopships of World War II*. US Army Center of Military History. Page 139. Retrieved from https://history.army.mil/documents/WWII/wwii_Troopships.pdf.

10. Shipscribe. Retrieved from http://www.shipscribe.com/usnaux/AP/AP63.html.

11. Charles. (1947). Page 139.

SS Monterey

12. Monterey. Retrieved from https://en.wikipedia.org/wiki/SS_Monterey.

13. Monterey. Retrieved from http://www.armed-guard.com/troop15.html.

USS Cepheus

14. Naval History and Heritage Command. Retrieved from https://www.history.navy.mil/research/histories/ship-histories/danfs/c/cepheus-i.html.

15. NavSource Online. (Cephus). Retrieved from http://www.navsource.org/archives/10/02/02018.htm.

USS Burleson

16. Naval History and Heritage Command. (Burleson). Retrieved from https://www.history.navy.mil/research/histories/ship-histories/danfs/b/burleson-i.html.

17. NavSource Online. (Burleson). Retrieved from https://www.navsource.org/archives/10/03/03067.htm.

18. Naval History and Heritage Command. (Burleson).

19. Naval History and Heritage Command. (Burleson).

Notes Appendix F. Airplanes Serviced by CASU-11

1. Naval History and Heritage Command. (1945). *Location of US Naval Aircraft - World War II, 2 February 1942 through 7 September 1945.* Retrieved from https://www.history.navy.mil/research/histories/naval-aviation-history/involvement-by-conflict/world-war-ii/location-of-us-naval-aircraft-world-war-ii.html.

The following books were used to assemble Appendix F;
- *The International Encyclopedia of Aircraft (IEOA). (*1991*).* United States of America: The Mallard Press.

- Ethell, J.L., Watanabe, R. (1996). *Great Book of World War II Airplanes.* Gramercy

- Angelucci, E., Matricardi, P. (2001). *Complete Book of World War II Combat Aircraft.* White Star

- Jackson, R. (2017). *Aircraft of World War II Development, Weaponry, Specifications.* Amber Books

Notes Appendix G. SBD Mission Planning and Flying

1. Intelligence, Strike Command, Commander Aircraft Solomons. (1944). *Facts of a Typical Strike.* Retrieved from https://www.fold3.com/image/251/276937000.

2. Watkinson, W. *Explains SBD Dive Brakes.* https://www.youtube.com/watch?v=KzptkuknMR4.

Notes Appendix H. After Action Report on CASUs

1. Chief of Naval Operations. (1944). Letter Op-33-SCR Serial: 0051333 0f 17 Jun 1944. Retrieved from https://www.fold3.com/image/302034003. Pages 1-2.

2. Commander South Pacific Area and Force. (1945). *History of Commander Aircraft South Pacific, Forwarding of.* Serial 156197 date 31

Dec 1945. Retrieved from https://www.fold3.com/image/1/302031415. Pages 10 - 12.

3. Commander South Pacific Area and Force. (1945). Page 14.

4. Commander South Pacific Area and Force. (1945). Page 29.

5. Commander South Pacific Area and Force. (1945). Pages 30-31.

6. Commander South Pacific Area and Force. (1945). Pages 146-149.

7. Commander South Pacific Area and Force. (1945). Pages 152-153.

8. Commanding Officer, Torpedo Squadron Twenty Eight letter of 29 Jan 1945, History of Torpedo Squadron Twenty Eight. https://www.fold3.com/image/1/302038266. Narrative, Pages 1 and 2.

9. Commander Patrol Bombing Squadron One Hundred Nine letter Serial 009 date unknown, Summary of Operations, Patrol Bombing Squadron One Hundred Nine. https://www.fold3.com/image/300888419. Page 6

10. Commander Patrol Bombing Squadron One Hundred Nine Ibid. Page 6. https://www.fold3.com/image/300888435

11. Commanding Officer, Patrol Bombing Squadron One Hundred Eighteen, letter Serial 014 of 25 Jul 1945, Summary of Operations, Patrol Bombing Squadron One Hundred Eighteen. https://www.fold3.com/image/251/300850418. Page 6.

Notes Appendix I. CASU-11 Chronology

1. No notes.

Bibliography

Websites

Fold3, https://www.fold3.com/ Access to military records from U.S. National Archives.

Digital Commons @ LibertyUniversity, https://digitalcommons.liberty.edu/cgi/viewcontent.cgi?referer=&httpsredir=1&article=1021&context=hist_fac_pubs

HyperWar, A HYPERTEXT history of the Second World War, http://www.ibiblio.org/hyperwar/

WikipediA, The Free Encyclopedia, https://en.wikipedia.org/wiki

Remembering Okinawa History, https://www.rememberingokinawa.com/page/remembering_okinawa_home

Naval History and Heritage Command, https://www.history.navy.mil/

Books

Astor, Gerald. *SEMPER FI IN THE SKY: The Marine Air Battles of World War II.* New York: Ballatine Books, 2005.

Ballantine, Duncan S. *U.S. Naval Logistics in the Second World War.* Newport: Naval War College Press, 1998.

Bibliography

Bennett, Judith A. *NATIVES and EXOTICS: World War II and Environment in the Southern Pacific.* Honolulu: University of Hawaii Press, 2009.

Bergerud, Eric M. *Fire in the Sky: The Air War in the South Pacific.* Boulder: Westview Press, 2000.

Boyington, Gregory Col. USMC, Ret. *BAA BAA BLACK SHEEP.* New York: Bantam Books, 1977.

Carter, Worrall R. *Beans, Bullets and Black Oil.* Newport: Naval War College Press, 1998.

Cortesi, Lawrence. *Operation Cartwheel: The Final Countdown to V-J Day.* New York: Kensington Publishing, 1982.

Davis, Donald A. *Lightning Strike.* New York: St. Martin's Press, 2005.

Feifer, George, *Tennozan: the Battle of Okinawa and the atomic bomb.* New York: Houghton Mifflin, 1992.

Foster, John M. Capt. USMCR, *Hell in the Heavens.* New York: G.P. Putnam's Sons, 1961.

Frank, Richard B. *Guadalcanal: The Definitive Account of the Landmark Battle.* New York: Random House, 1990.

Fussell, Paul. *WARTIME: Understanding and Behavior in the Second World War.* New York: Oxford University Press, 1989.

Gilbert, Alton Keith. *A Leader Born: The Life of Admiral John Sidney McCain, Pacific Carrier Commander.* Philadelphia: Casemate, 2006.

Gailey, Harry A. *The War in the Pacific - From Pearl Harbor to Tokyo Bay.* Navato: Presidio Press 1995.

Hagen, Jerome T. *War in the Pacific, America at War.* Vol I. Honolulu: Hawaii Pacific University, 2006.

Bibliography

Jones, James. *The Thin Red Line.* New York: Dell Publishing, 1962.

Lane, Kerry L. *GuadalcanalMarine.* Jackson, MS: University Press of Mississippi. Page 186.

Leckie, Robert. *Helmet for my Pillow.* New York: Simon & Schuster, Inc., 2001.

McGee, William L. *Amphibious Operations in the South Pacific in WWII. Vol II: THE SOLOMONS CAMPAIGNS 1942-1943 From Guadalcanal to Bougainville PACIFIC WAR TURNING POINT.* Santa Barbara: BMC Publications, 2002.

Merillat, Herbert Christian. *Guadalcanal Remembered.* New York: Dodd, Mead & Company, 1982.

Mersky, Peter B. *U.S. MARINE CORPS AVIATION 1912 to the Present.* Baltimore: The Nautical & Aviation Publishing Co., 1983.

Miller, Vern A. *Our Coral Carriers Helped Turn the Tide of Battle,* http://www.casu44.com/casu.html

Moore, Stephen L. *PACIFIC PAYBACK: The Carrier Aviators Who Avenged Pearl Harbor at the Battle of Midway.* New York: NAI Caliber, 2014.

Morison, Samuel Eliot. *History of United States Naval Operations in World War II. Vol. VI: Breaking the Bismarcks Barrier, 22 July 1942 - 1 May 1944.* Chicago: University of Illinois Press, 1978.

Morison, Samuel Eliot. *History of United States Naval Operations in World War II. Vol. XIV: Victory in the Pacific 1945.* Boston: Little Brown & Co., 1990.

Morison, Samuel Eliot. *History of United States Naval Operations in World War II. Vol. V: The Struggle for Guadalcanal, August 1942 - February 1943.* Annapolis: Naval Institute Press, 1949.

Bibliography

Morison, Samuel Eliot. *History of United States Naval Operations in World War II. Vol. XV: Supplement and General Index*. Boston: Little Brown & Co., 1984.

Potter, E.B., and Chester Nimitz. *SEAPOWER; A Naval History*. Englewood Cliffs: Prentice - Hall, 1960.

Prados, John. *Islands of Destiny*. New York: NAI Caliber, 2012.

Sherrod, Robert. *History of Marine Corps Aviation in World War II*. Baltimore, MD: Nautical & Aviation Publishing Company of America, 1987.

Sledge, E. B. *With the Old Breed at Peleliu and Okinawa*. New York: Oxford University Press, 1981.

Taylor, Theodore. *The Magnificent Mitscher*. Annapolis: Naval Institute Press, 1991.

The Aviation History Unit OP-519B, DCNO (Air) Editor, Buchanan, A. R. *THE NAVY'S AIR WAR: A Mission Completed*. New York: Harper & Brothers Publishers, 1946.

The United States Air Force Special Studies. *Air Warfare and Air Base Air Defense*. Washington, D.C.: Government Printing Office, 1988.

U.S. Army in World War II. *Cartwheel: The Reduction of RABAUL*. Washington, D.C.: Center of Military History, 1990.

U.S. Army in World War II. *Guadalcanal: The First Offensive*. Washington, D.C.: Center of Military History, 1989.

U.S. Army in World War II. *Okinawa: The Last Battle*. Washington, D.C.: Center of Military History, 1993.

Wolfert, Ira. *Battle for the Solomons*. Boston: Houghton Mifflin Company, 1943.

Wukovits, John F. *Black Sheep*. Annapolis: Naval Institute Press, 2011.